Kirby Deater-Deckard
Stephen A. Petrill
Editors

Gene-Environment Processes in Social Behaviors and Relationships

Gene-Environment Processes in Social Behaviors and Relationships has been co-published simultaneously as *Marriage & Family Review*, Volume 33, Numbers 1/2/3 2001.

Pre-publication
REVIEWS,
COMMENTARIES,
EVALUATIONS . . .

"During recent years there have been somewhat fruitless battles on whether family influences or peer influences are more important in children's psychological development. THIS BOOK IS BOTH INNOVATIVE AND HELPFUL in seeking to bring the two sets of influences together through a range of studies using twin, adoptee, and stepfamily designs to assess how genetic and environmental influences may work together in bringing about individual differences in children's emotions, behavior and especially social relationships. The different research approaches provide some new ways of thinking about, and investigating, how interpersonal relationships develop and have their effects."

Michael Rutter, MD, FRS
Professor of Developmental Psychopathology,
Institute of Psychiatry,
King's College, London

Gene-Environment Processes in Social Behaviors and Relationships

Gene-Environment Processes in Social Behaviors and Relationships has been co-published simultaneously as *Marriage & Family Review*, Volume 33, Numbers 1/2/3 2001.

The *Marriage & Family Review* Monographic "Separates"

Below is a list of "separates," which in serials librarianship means a special issue simultaneously published as a special journal issue or double-issue *and* as a "separate" hardbound monograph. (This is a format which we also call a "DocuSerial.")

"Separates" are published because specialized libraries or professionals may wish to purchase a specific thematic issue by itself in a format which can be separately cataloged and shelved, as opposed to purchasing the journal on an on-going basis. Faculty members may also more easily consider a "separate" for classroom adoption.

"Separates" are carefully classified separately with the major book jobbers so that the journal tie-in can be noted on new book order slips to avoid duplicate purchasing.

You may wish to visit Haworth's website at . . .

http://www.HaworthPress.com

. . . to search our online catalog for complete tables of contents of these separates and related publications.

You may also call 1-800-HAWORTH (outside US/Canada: 607-722-5857), or Fax 1-800-895-0582 (outside US/Canada: 607-771-0012), or e-mail at:

getinfo@haworthpressinc.com

Gene-Environment Processes in Social Behaviors and Relationships, edited by Kirby Deater-Deckard, PhD, and Stephen A. Petrill, PhD (Vol. 33, No. 1/2/3, 2001). *"During recent years there have been somewhat fruitless battles on whether family influences or peer influences are more important in children's psychological development. This book is both innovative and helpful in seeking to bring the two sets of influences together through a range of studies using twin, adoptee, and stepfamily designs to assess how genetic and environmental influences may work together in bringing about individual differences in children's emotions, behavior and especially social relationships. The different research approaches provide some new ways of thinking about, and investigating, how interpersonal relationships develop and have their effects." (Michael Rutter, MD, FRS, Professor of Developmental Psychopathology, Institute of Psychiatry, King's College, London)*

Pioneering Paths in the Study of Families: The Lives and Careers of Family Scholars, edited by Suzanne K. Steinmetz, PhD, MSW, and Gary W. Peterson, PhD (Vol. 30, No. 3, 2000; Vol. 30, No. 4, 2001; Vol. 31, No. 1/2/3/4; Vol. 32, No. 1/2, 2001).

FATHERHOOD: Research, Interventions and Policies, edited by H. Elizabeth Peters, PhD, Gary W. Peterson, PhD, Suzanne K. Steinmetz, PhD, MSW, and Randal D. Day, PhD (Vol. 29, No. 2/3/4, 2000). *Brings together the latest facts to help researchers explore the father-child relationship and determine what factors lead fathers to be more or less involved in the lives of their children, including human social behavior, not living with a child, being denied visiting privileges, and social norms regarding gender differences versus work responsibilities.*

Concepts and Definitions of Family for the 21st Century, edited by Barbara H. Settles, PhD, Suzanne K. Steinmetz, PhD, MSW, Gary W. Peterson, PhD, and Marvin B. Sussman, PhD (Vol. 28, No. 3/4, 1999). *Views family from a U.S. perspective and from many different cultures and societies. The controversial question "What is family?" is thoroughly examined as it has become an increasingly important social policy concern in recent years as the traditional family has changed.*

The Role of the Hospitality Industry in the Lives of Individuals and Families, edited by Pamela R. Cummings, PhD, Francis A. Kwansa, PhD, and Marvin B. Sussman, PhD (Vol. 28, No. 1/2, 1998). *"A must for human resource directors and hospitality educators." (Dr. Lynn Huffman, Director, Restaurant, Hotel, and Institutional Management, Texas Tech University, Lubbock, Texas)*

Stepfamilies: History, Research, and Policy, edited by Irene Levin, PhD, and Marvin B. Sussman, PhD (Vol. 26, No. 1/2/3/4, 1997). *"A wide range of individually valuable and stimulating chapters that form a wonderfully rich menu from which readers of many different kinds will find exciting and satisfying selections." (Jon Bernardes, PhD, Principal Lecturer in Sociology, University of Wolverhampton, Castle View Dudley, United Kingdom)*

Families and Adoption, edited by Harriet E. Gross, PhD, and Marvin B. Sussman, PhD (Vol. 25, No. 1/2/3/4, 1997). *"Written in a lucid and easy-to-read style, this volume will make an invaluable contribution to the adoption literature." (Paul Sachdev, PhD, Professor, School of Social Work, Memorial University of Newfoundland, St. John's, Newfoundland, Canada)*

The Methods and Methodologies of Qualitative Family Research, edited by Jane F. Gilgun, PhD, LICSW, and Marvin B. Sussman, PhD (Vol 24, No. 1/2/3/4, 1997). *"An authoritative look at the usefulness of qualitative research methods to the family scholar." (Family Relations)*

Intercultural Variation in Family Research and Theory: Implications for Cross-National Studies, Volumes I and II, edited by Marvin B. Sussman, PhD, and Roma S. Hanks, PhD (Vol. 22, No. 1/2/3/4, and Vol. 23, No. 1/2/3/4, 1997). *Documents the development of family research in theory in societies around the world and inspires continued cross-national collaboration on current research topics.*

Families and Law, edited by Lisa J. McIntyre, PhD, and Marvin B. Sussman, PhD (Vol. 21, No. 3/4, 1995). *With this new volume, family practitioners and scholars can begin to increase the family's position in relation to the law and legal system.*

Exemplary Social Intervention Programs for Members and Their Families, edited by David Guttmann, DSW, and Marvin B. Sussman, PhD (Vol. 21, No. 1/2, 1995). *An eye-opening look at organizations and individuals who have created model family programs that bring desired results.*

Single Parent Families: Diversity, Myths and Realities, edited by Shirley M. H. Hanson, RN, PhD, Marsha L. Heims, RN, EdD, Doris J. Julian, RN, EdD, and Marvin B. Sussman, PhD (Vol. 20, No. 1/2/3/4, 1994). *"Remarkable! . . . A significant work and is important reading for multidisciplinary family professionals including sociologists, educators, health care professionals, and policymakers." (Maureen Leahey, RN, PhD, Director, Outpatient Mental Health Program, Director, Family Therapy Training Program, Calgary District Hospital Group)*

Families on the Move: Immigration, Migration, and Mobility, edited by Barbara H. Settles, PhD, Daniel E. Hanks III, MS, and Marvin B. Sussman, PhD (Vol 19, No 1/2/3/4, 1993). *Examines the current research on family mobility, migration, and immigration and discovers new directions for understanding the relationship between mobility and family life.*

American Families and the Future: Analyses of Possible Destinies, edited by Barbara H. Settles, PhD, Roma S. Hanks, PhD, and Marvin B. Sussman, PhD (Vol. 18, No. 3/4, 1993). *This book discusses a variety of issues that face and will continue to face families in coming years and describes various strategies families can use in their decision-making processes.*

Publishing in Journals on the Family: Essays on Publishing, edited by Roma S. Hanks, PhD, Linda Matocha, PhD, RN, and Marvin B. Sussman, PhD (Vol. 18, No. 1/2, 1993). *This helpful book contains varied perspectives from scholars at different career stages and from editors of major publication outlets, providing readers with important information necessary to help them systematically plan a productive scholarly career.*

Publishing in Journals on the Family: A Survey and Guide for Scholars, Practitioners, and Students, edited by Roma S. Hanks, PhD, Linda Matocha, PhD, RN, and Marvin B. Sussman, PhD (Vol. 17, No. 3/4, 1992). *"Comprehensive. . . . Includes listings for some 200 social science journals whose editors have expressed an interest in publishing empirical research and theoretical articles about the family." (Reference & Research Book News)*

Wider Families: New Traditional Family Forms, edited by Teresa D. Marciano, PhD, and Marvin B. Sussman, PhD (Vol. 17, No. 1/2, 1992). *"An insightful and informative compilation of essays on the subject of wider families." (Journal of Marriage and the Family)*

Families: Intergenerational and Generational Connections, edited by Susan K. Pfeifer, PhD, and Marvin B. Sussman, PhD (Vol. 16, No. 1/2/3/4, 1991). *"The contributors challenge and move dramatically from outdated myths and stereotypes concerning who and what is family, what its members do, and how they continue its traditions to contemporary views of families and their relationships." (Contemporary Psychology)*

Corporations, Businesses, and Families, edited by Roma S. Hanks, PhD, and Marvin B. Sussman, PhD (Vol. 15, No. 3/4, 1991). *"Examines the changing relationship between family systems and work organizations." (Economic Books)*

Families in Community Settings: Interdisciplinary Perspectives, edited by Donald G. Unger, PhD, and Marvin B. Sussman, PhD (Vol. 15, No. 1/2, 1990). *"An excellent introduction in which to frame and understand the central issues." (Abraham Wandersman, PhD, Professor, Department of Psychology, University of South Carolina)*

Homosexuality and Family Relations, edited by Frederick W. Bozett, RN, DNS, and Marvin B. Sussman, PhD (Vol. 14, No. 3/4, 1990). *"Offers a smorgasbord of familial topics. . . . Provides references for those seeking more information." (Lesbian News)*

Cross-Cultural Perspectives on Families, Work, and Change, edited by Katja Boh, PhD, Giovanni Sgritta, PhD, and Marvin B. Sussman, PhD (Vol. 14, No. 1/2, 1990). *"On the cutting edge of this new perspective that sees a modern society as a set of influences that affect human beings and not just a collection of individual orphans." (John Mogey, DSc, Adjunct Professor of Sociology, Arizona State University)*

Museum Visits and Activities for Family Life Enrichment, edited by Barbara H. Butler, PhD, and Marvin B. Sussman, PhD (Vol. 13, No. 3/4, 1989). *"Very interesting reading . . . a fine synthesis of current thinking concerning families in museums." (Jane R. Glaser, Special Assistant, Office of the Assistant Secretary for Museums, Smithsonian Institution, Washington, DC)*

AIDS and Families, edited by Eleanor D. Macklin, PhD (Vol. 13, No. 1/2, 1989). *"A highly recommended book. Will provide family professionals, policymakers, and researchers with a foundation for further exploration on the largely unresearched topic of AIDS and the family." (Family Relations)*

Transitions to Parenthood, edited by Rob Palkovitz, PhD, and Marvin B. Sussman, PhD (Vol. 12, No. 3/4, 1989). *In this insightful volume, experts discuss the issues, changes, and problems involved in becoming a parent.*

Deviance and the Family, edited by Frank E. Hagan, PhD, and Marvin B. Sussman, PhD (Vol. 12, No. 1/2, 1988). *Leading experts in the fields of criminal justice, sociology, and family services explain the causes of deviance as well as the role of the family.*

Alternative Health Maintenance and Healing Systems for Families, edited by Doris Y. Wilkinson, PhD, and Marvin B. Sussman, PhD (Vol. 11, No. 3/4, 1988). *This important book offers timely discussions of current approaches and treatments in modern medicine that have had great impact upon family health care.*

'Til Death Do Us Part: How Couples Stay Together, edited by Jeanette C. Lauer and Robert C. Lauer (Supp. #1, 1987). *"A landmark study that will serve as a classic for the emerging ethic of commitment to marriage, family, and community." (Gregory W. Brock, PhD, Professor of Family Science and Marriage and Family Therapy, University of Wisconsin)*

Childhood Disability and Family Systems, edited by Michael Ferrari, PhD, and Marvin B. Sussman, PhD (Vol. 11, No. 1/2, 1987). *A motivating book that offers new and enlightening perspectives for professionals working with disabled children and their families.*

Family Medicine: The Maturing of a Discipline, edited by William J. Doherty, PhD, Charles E. Christianson, MD, ScM, and Marvin B. Sussman, PhD (Vol. 10, No. 3/4, 1987). *"Well-written essays and a superb introduction concerning various aspects of the field of family medicine (or as it is sometimes called, family practice)." (The American Journal of Family Therapy)*

Families and the Prospect of Nuclear Attack/Holocaust, edited by Teresa D. Marciano, PhD, and Marvin B. Sussman, PhD (Vol. 10, No. 2, 1986). *Experts address the issues and effects of the continuing threat of nuclear holocaust on the behavior of families.*

The Charybdis Complex: Redemption of Rejected Marriage and Family Journal Articles, edited by Marvin B. Sussman, PhD (Vol. 10, No. 1, 1986). *An examination of the "publish-or-perish" syndrome of academic publishing, with a frank look at peer review.*

Men's Changing Roles in the Family, edited by Robert A. Lewis, PhD, and Marvin B. Sussman, PhD (Vol. 9, No. 3/4, 1986). *"Brings together a wealth of findings on men's family role enactment . . . provides a well-integrated, carefully documented summary of the literature on men's roles in the family that should be useful to both family scholars (in their own work and the classroom) and practitioners." (Contemporary Sociology)*

Families and the Energy Transition, edited by John Byrne, David A. Schulz, and Marvin B. Sussman, PhD (Vol. 9, No. 1/2, 1985). *An important appraisal of the future of energy consumption by families and the family's adaptations to decreasing energy availability.*

Pets and the Family, edited by Marvin B. Sussman, PhD (Vol. 8, No. 3/4, 1985). *"Informative and thorough coverage of what is currently known about the animal/human bond." (Canada's Mental Health)*

Personal Computers and the Family, edited by Marvin B. Sussman, PhD (Vol 8, No. 1/2, 1985). *A pioneering volume that explores the impact of the personal computer on the modern family.*

Women and the Family: Two Decades of Change, edited by Beth B. Hess, PhD, and Marvin B. Sussman, PhD (Vol. 7, No. 3/4, 1984). *"A scholarly, thorough, readable, informative, well-integrated, current overview of social science research on women and the family." (Journal of Gerontology)*

Obesity and the Family, edited by David J. Kallen, PhD, and Marvin B. Sussman, PhD (Vol. 7, No. 1/2, 1984). *"Should be required reading for all persons touched by the problem of obesity–the teachers, the practitioners of every discipline, and the obese themselves." (Journal of Nutrition Education)*

Human Sexuality and the Family, edited by James W. Maddock, PhD, Gerhard Neubeck, EdD, and Marvin B. Sussman, PhD (Vol. 6, No. 3/4, 1984). *"Twelve chapters that not only add some new ideas about the place of sexuality in the family but also go beyond this to show how widely sexuality influences human behavior and thought . . . excellent." (Siecus Report)*

Social Stress and the Family: Advances and Developments in Family Stress Theory and Research, edited by Hamilton I. McCubbin, Marvin B. Sussman, PhD, and Joan M. Patterson (Vol. 6, No. 1/2, 1983). *An informative anthology of recent theory and research developments pertinent to family stress.*

The Ties That Bind: Men's and Women's Social Networks, edited by Laura Lein, PhD, and Marvin B. Sussman, PhD (Vol. 5, No. 4, 1983). *An examination of the networks for men and women in a variety of social contexts.*

Family Systems and Inheritance Patterns, edited by Judith N. Cates and Marvin B. Sussman, PhD (Vol. 5, No. 3, 1983). *Specialists in economics, law, psychology, and sociology provide a comprehensive examination of the disposition of property following a death.*

Alternatives to Traditional Family Living, edited by Harriet Gross, PhD, and Marvin B. Sussman, PhD (Vol. 5, No. 2, 1982). *"Professionals interested in the lifestyles described will find well-written essays on these topics." (The Amercian Journal of Family Therapy)*

Intermarriage in the United States, edited by Gary A. Crester, PhD, and Joseph J. Leon, PhD (Vol. 5, No. 1, 1982). *"A very good compendium of knowledge and of theoretical and technical issues in the study of intermarriage." (Journal of Comparative Family Studies)*

Monographs "Separates" list continued at the back

Gene-Environment Processes in Social Behaviors and Relationships

Kirby Deater-Deckard
Stephen A. Petrill
Editors

Gene-Environment Processes in Social Behaviors and Relationships has been co-published simultaneously as *Marriage & Family Review*, Volume 33, Numbers 1/2/3 2001.

The Haworth Press, Inc.
New York • London • Oxford

Gene-Environment Processes in Social Behaviors and Relationships
has been co-published simultaneously as *Marriage & Family Review*™,
Volume 33, Numbers 1/2/3 2001.

The development, preparation, and publication of this work has been undertaken with great care. However, the publisher, employees, editors, and agents of The Haworth Press and all imprints of The
Haworth Press, Inc., including The Haworth Medical Press® and Pharmaceutical Products Press®, are
not responsible for any errors contained herein or for consequences that may ensue from use of materials or information contained in this work. Opinions expressed by the author(s) are not necessarily those
of The Haworth Press, Inc. With regard to case studies, identities and circumstances of individuals discussed herein have been changed to protect confidentiality. Any resemblance to actual persons, living
or dead, is entirely coincidental.

Cover design by MaryLouise Doyle

Library of Congress Cataloging-in-Publication Data

Gene-environment processes in social behaviors and relationships / Kirby Deater-Deckard, Ste-
phen A. Petrill, editors.
    p. cm.
"Co-published simultaneously as Marriage & family review, Volume 33, Numbers 1/2/3 2001."
Includes bibliographical references and index.
    ISBN 0-7890-1956-6 (hard : alk. paper) – ISBN 0-7890-1957-4 (pbk : alk. paper)
1. Nature and nurture. 2. Individual differences in children. 3. Interpersonal relations in children.
I. Deater-Deckard, Kirby D. II. Petrill, Stephen A., 1968- III. Marriage & family review.
    BF341 .G38 2002
    155.7–dc21
                                                                                    2002014183

Indexing, Abstracting & Website/Internet Coverage

This section provides you with a list of major indexing & abstracting services. That is to say, each service began covering this periodical during the year noted in the right column. Most Websites which are listed below have indicated that they will either post, disseminate, compile, archive, cite or alert their own Website users with research-based content from this work. (This list is as current as the copyright date of this publication.)

Abstracting, Website/Indexing Coverage . . . . Year When Coverage Began

- *Abstracts in Social Gerontology: Current Literature on Aging* **1993**

- *Academic Abstracts/CD-ROM* **1993**

- *Academic ASAP <www.galegroup.com>* **1989**

- *Academic Search: Database of 2,000 selected academic serials, updated monthly: EBSCO Publishing* **1996**

- *Academic Search Elite (EBSCO)* **1993**

- *AgeLine Database* **1991**

- *AGRICOLA Database <www.natl.usda.gov/ag98>* **1992**

- *Applied Social Sciences Index & Abstracts (ASSIA) (Online: ASSI via Data-Star) (CDRom: ASSIA Plus) <www.csa.com>* **1993**

- *CNPIEC Reference Guide: Chinese National Directory of Foreign Periodicals* **1996**

(continued)

(continued)

Special Bibliographic Notes related to special journal issues (separates) and indexing/abstracting:

- indexing/abstracting services in this list will also cover material in any "separate" that is co-published simultaneously with Haworth's special thematic journal issue or DocuSerial. Indexing/abstracting usually covers material at the article/chapter level.
- monographic co-editions are intended for either non-subscribers or libraries which intend to purchase a second copy for their circulating collections.
- monographic co-editions are reported to all jobbers/wholesalers/approval plans. The source journal is listed as the "series" to assist the prevention of duplicate purchasing in the same manner utilized for books-in-series.
- to facilitate user/access services all indexing/abstracting services are encouraged to utilize the co-indexing entry note indicated at the bottom of the first page of each article/chapter/contribution.
- this is intended to assist a library user of any reference tool (whether print, electronic, online, or CD-ROM) to locate the monographic version if the library has purchased this version but not a subscription to the source journal.
- individual articles/chapters in any Haworth publication are also available through the Haworth Document Delivery Service (HDDS).

ABOUT THE EDITORS

Kirby Deater-Deckard, PhD, is Associate Professor in the Department of Psychology at the University of Oregon in Eugene. He taught and conducted research previously at Vanderbilt University and universities in the United Kingdom and Sweden. Dr. Deater-Deckard has published extensively and has also served as a reviewer for several professional journals, including *Child Development, Cognition and Emotion, Current Directions in Psychological Science, Social Development,* the *British Journal of Developmental Psychology,* and the *Journal of Family Psychology.* He has presented at a number of conferences in North America and Europe.

Stephen A. Petrill, PhD, is Assistant Professor in the Center for Developmental and Health Genetics at the Pennsylvania State University in University Park. He taught previously at Wesleyan University, Case Western Reserve University, Ashland University, and the Hawken Upper Preparatory School. He has published many journal articles and book chapters, and has served as a reviewer for several professional journals, including *Behavior Genetics,* the *British Journal of Clinical Psychology, Current Directions in Psychological Science, School Psychology Review, Intelligence,* and the *Journal of Personality and Social Psychology.* Dr. Petrill has been a presenter at more than two dozen national and international conferences and has been the recipient of the International Society for the Study of Individual Differences Early Career Award and the Award for Excellence in Research from the American Mensa Education and Research Foundation.

Gene-Environment Processes in Social Behaviors and Relationships

CONTENTS

MARRIAGE & FAMILY REVIEW
EDITORS' COMMENT

The Nature-Nurture controversy, stated as such, has been around for at least a century. It has been used to describe specific populations, provide educational programs, and place children for adoption, to name just a few instances. Freud's influence, specifically his conclusion that "biology is destiny," greatly strengthened this position. There are numerous instances of children from impoverished, uneducated, or African American backgrounds being sterilized so that they would not be able to reproduce similarly "damaged" children. However, behaviorism, under Watson's influence, gave rise to the notion that the environment played a more critical role.

This stance was strengthened during the post-World War II environment, when any idea of racial or genetic reasons for denying opportunities or making life-altering decisions was met with horror. The Civil Rights movement gave voice to the idea that race was not the reason for African Americans' making less progress relative to whites; it was discrimination and lack of opportunities. The women's movement furthers this position. There was strong support that our destiny was controlled by our environment and with the proper nurturance, adversity could be overcome.

Research, specifically cognitive mapping and studies of brain activity during the 1980s, suggested that gender-based differences could be observed in different areas of brain activity. These findings suggested that biology might play a greater role in destiny. Studies of difficult

[Haworth co-indexing entry note]: *"Marriage & Family Review* Editors' Comment." Steinmetz, Suzanne K., and Gary W. Peterson. Co-published simultaneously in *Marriage & Family Review* (The Haworth Press, Inc.) Vol. 33, No. 1, 2001, pp. 1-2; and: *Gene-Environment Processes in Social Behaviors and Relationships* (ed: Kirby Deater-Deckard, and Stephen A. Petrill) The Haworth Press, Inc., 2001, pp. 1-2. Single or multiple copies of this article are available for a fee from The Haworth Document Delivery Service [1-800-HAWORTH, 9:00 a.m. - 5:00 p.m. (EST). E-mail address: getinfo@haworthpressinc.com].

adoption provided additional support when they revealed that even the best environment might not erase nature-based problems. We learned that many mental health problems appeared to have a genetic basis and were not totally the result of poor parenting or a bad environment.

In the age of DNA, we now have a powerful tool to identify more accurately the role that nature and nurture play in individual development. The papers in this collection provide ample evidence to help us more clearly identify the specific role that nature and nurture play.

Suzanne K. Steinmetz
Gary W. Peterson

Examining Social Behavior and Relationships Using Genetically Sensitive Designs: An Introduction

Stephen A. Petrill

Over the past 20 years, it has become widely accepted that both genes and environments are important when examining outcome variables such as temperament, behavior problems, and cognitive ability (Plomin, DeFries, McClearn, & McGuffin, 2000). However, the implications of these well-replicated findings have not been fully integrated into theories examining dyadic and triadic processes within families or within peer groups. If we assume that differences in the way people behave are influenced, in part, by genetic factors, we should also assume that genetic influences may affect the relationships among family and peer members. The purpose of this volume is to describe a set of studies that demonstrate not only how genetically sensitive designs may be used to examine family/peer processes and outcomes, but also why such studies are important to theories of social development. The purpose of this introduction is to describe the logic behind the various methods employed by the authors and to introduce the manuscripts that compose the collection. More detailed discussions of the statistical procedures by

Stephen A. Petrill is affiliated with the Center for Developmental and Health Genetics, Department of Biobehavioral Health, Pennsylvania State University, 101 Amy Gardner House, University Park, PA 16802.

[Haworth co-indexing entry note]: "Examining Social Behavior and Relationships Using Genetically Sensitive Designs: An Introduction." Petrill, Stephen A. Co-published simultaneously in *Marriage & Family Review* (The Haworth Press, Inc.) Vol. 33, No. 1, 2001, pp. 3-10; and: *Gene-Environment Processes in Social Behaviors and Relationships* (ed: Kirby Deater-Deckard, and Stephen A. Petrill) The Haworth Press, Inc., 2001, pp. 3-10. Single or multiple copies of this article are available for a fee from The Haworth Document Delivery Service [1-800-HAWORTH, 9:00 a.m. - 5:00 p.m. (EST). E-mail address: getinfo@haworthpressinc.com].

3

which these methods are conducted in practice may be found elsewhere (Neale & Cardon, 1992; Neale, 1995).

EXAMINING GENETIC
AND ENVIRONMENTAL INFLUENCES

Determining Genetic Influences

Behavioral genetics methods are based on the assumption that individual differences in a measured behavior may be decomposed into discrete sources of genetic and environmental variance. This is accomplished by examining family members who possess different levels of genetic relatedness. For example, identical twins share 100% of the same genes while fraternal twins share 50% of the same genes, on average. Therefore, if identical twins are more similar than fraternal twins, then genetic influences are implicated, presumably because identical twins are more similar genetically than fraternal twins. In a similar vein, if adopted children are more similar to their biological parents or biological siblings than their adoptive parents and siblings, then genetic influences are also implicated. Heritability (h^2) is the statistic that measures the extent to which individual differences on an outcome measure are influenced by genetic differences in the population.

Describing Environmental Influences

Behavioral genetic methods also provide a powerful tool to study the environment. Using special populations such as twins and adoptees, the environment can be divided into two types. The first is called the shared environment (or c^2). Shared Environment is defined as ANY environment that makes family members similar. These influences can occur both *inside* and/or *outside* of the home. For example, two siblings growing up together live in the same home, go to the same school, or may have some of the same friends. In twin studies, shared environment is suggested when identical twins are no more similar than fraternal twins, presumably because it is the shared environment that is causing the twins to be similar. It is assumed that this shared environment is experienced similarly in identical and fraternal twins. In adoption designs, shared environment is implicated when significant relationships are found between family members who are genetically unrelated. In this case, it can only be the shared environment that makes unrelated family members similar.

Finally, behavioral genetic methods may be used to quantify the nonshared environment (or e^2). The nonshared environment is anything that makes family members different. Although siblings may have the same parents, those parents may treat the siblings differently. Although siblings may go to the same school, they may have different friends or different educational experiences. The nonshared environment is best measured by examining the differences between identical twins living together. Because they share 100% of the same genes and live in the same home, any differences between identical twins living together must be due to two sources of variance–nonshared environment and error.

Beyond Heritability

To this point, I have described behavioral genetics as a univariate methodology: Variance in a behavior may be decomposed into genetic, shared environmental, or nonshared environmental influences. However, while knowing something about the genetic and environmental influences on individual measures is interesting, for behavioral genetic methods to have theoretical meaning they must also describe how genes and environments influence the relationships among variables. Multivariate behavioral genetic methods are based on the cross-twin, or cross-family correlation. Say, for example, that we compare twin 1's score on anxiety and twin 2's score on depression. If identical twins are more similar than fraternal twins or biologically related siblings are more similar than adoptive siblings, then it is assumed that genes have something to do with the relationship between anxiety and depression. If identical twins or biological siblings are no more similar than fraternal twins or adoptive siblings respectively, then shared environment is assumed to have an influence on the relationship between anxiety and depression. Finally, if anxiety and depression are correlated within a person but are not correlated across family members, then nonshared environmental influences are implicated. For example, in the case of identical twins, twin 1's anxiety and depression scores will correlate but twin 1's anxiety will not predict twin 2's depression.

Expanding from this simple bivariate case, the same logic may be applied to a set of variables. In this way, behavioral genetics methods may be used to ask theoretically meaningful questions about how a set of variables relates to one another. These variables may be assessed concurrently to examine the relationship among variables at a given age, or may be assessed longitudinally in order to study genetic and environmental effects upon development. Thus, any theoretically meaningful relationship among variables may be decomposed into its genetic and environmental components given the proper sample, behavioral measurement, and statistical approach.

Behavioral Genetics at the Extremes

The methods outlined thus far may be used to decompose individual differences in the unselected population into genetic and environmental components of variance. In other words, these methods can be used to examine how variance in depressive symptoms (both clinical and nonclinical) relates to anxiety (both high and low). However, in some cases it may be more useful to examine the genetic and environmental influences on diagnosed depression, or anxiety, or low cognitive ability. There are three main methods that have been used to describe genetic and environmental influences at the extremes:

a. *Concordance rates.* These statistics measure the proportion where both members of a family are selected into a group. If twin 1 is selected as having behavior problems, is twin 2 also likely to be selected as having behavior problems? The extent to which identical twins are more likely to both be selected than fraternal twins suggests genetic influences.

b. *Tetrachoric correlations.* One problem with concordance rates is that they are blind to the base rate of a disorder in the population. In other words, the importance of an identical twins concordance of 60% is more meaningful if the base rate of a disorder in the population is 5% as opposed to 50%. In essence, these correlations reconstruct the entire distribution of scores using the concordance rate and the base rate in the population

c. *DeFries-Fulker (DF) multiple regression analyses* (DeFries & Fulker, 1985; 1988). This approach allows for the selection of extreme groups while maintaining the quantitative distribution of scores. A cutoff score is set for a particular behavior and twins (or adoptees) are selected. The mean of this selected group is then calculated. The means of their siblings are then calculated. If there is no sibling relationship, then the mean of the siblings should be equal to the unselected population. If the sibling mean for the identical twins is closer to the cutoff than the sibling mean for fraternal twins, then genetic influences are implicated. The main advantage of this approach is that the sibling's scores can fall either inside or outside of the cutoff and still be counted, as opposed to concordance and tetrachoric-based approaches.

Gene-Environment Processes

So far, we have implicitly assumed that genes and environments are, in essence, main effects. The total variance in some outcome is a linear function of h^2, c^2, and e^2. However, it is also possible that genetic and environmental influences may correlate with one another. For example, a highly intelligent child will likely be living in a home with highly intelligent parents, who will provide more books in the home. Additionally, this highly intelligent child is likely to be tracked into a different educational path and may be more likely to seek out more intellectually challenging activities and peers than less intelligent children. Since genes account for roughly 1/2 the variance in cognitive ability, then it is likely that the probability of experiencing intellectually enriched environments is, in part, a function of the genes important to cognitive ability.

The same may be true when examining social relationships. Each of the studies presented as part of this volume examines how genetically sensitive designs may be used to understand gene-environment processes. Each of the papers assumes that, because genetic influences are significant in social behavior, it is necessary to employ genetically sensitive designs to understand social relationships. Each study employs a battery of the various quantitative genetic methods described above to address several important questions concerning the development of social relationships.

ARTICLES IN THIS COLLECTION

Conceptual Papers

There are three articles in this collection that are best described as conceptual papers. Towers, Spotts, and Neiderhiser describe the literature examining gene-environment processes in mother-child relationships. This manuscript provides further evidence that it is necessary to employ genetically sensitive designs to study the environment and social relationships.

Examining the issues surrounding the nonshared environment is the focus of the article by McGuire. As stated earlier in this introduction, the nonshared environment constitutes the most consistent source of environmental variance, at least from a developmental perspective. McGuire outlines the various methodologies and issues surrounding this strain of quantitative genetic research.

Although heritability estimates are useful, they do not identify the genes themselves nor does heritability describe the biological processes through which genes affect behavior. Over the past decade, genetic research has begun to identify and replicate stretches of DNA that impact behavior. Eley's article describes the various methodologies used to identify DNA markers important to social behavior and social development. Such molecular genetics work is important because it frees the researcher from having to employ special populations such as twins and families. Because these methods may be employed using any population, there has been a proliferation in the number of studies using a molecular genetics component. More importantly, these methods provide a means to better understand the biology of social development and social behavior.

Empirical Papers Examining Family Relationships

Several papers examine genetic and environmental influences on family relationships. In general, these papers examine the relationship between child-based behavior (e.g., personality, temperament, behavior problems, psychopathology) and the relationships these children share with their parents and siblings. These studies are described briefly below.

Using twin design, Lemery and Goldsmith examined the relationship between difficult temperament, parental warmth and control, and sibling cooperation and conflict in 3-8 year-olds. Interestingly, the results of this study suggest that although genetic influences associated with difficult temperament predicted sibling conflict, parental warmth and parental control were more predictive of the sibling relationship.

In a similar vein, Riggins-Caspers and Cadoret's study employed an adoption design to explore whether adoptive parenting behaviors can provide a protective factor for children who are biologically at risk for psychopathology. Adoptive children were deemed to be at biological risk for psychopathology if the birth parent was diagnosed with psychopathology using *DSM-IIIR* criteria. The results of this study suggest that greater maternal warmth and lesser maternal overprotection reduced problem behaviors in adolescent adoptees who are biologically at risk for psychopathology.

Another important issue raised in the behavioral genetic literature is the extent to which family processes in twin and adoptive families are generalizable to other families. In Part II, the article by Deater-Deckard, Dunn, O'Connor, Davies, and Golding examines mother, partner (the child's parent's partner), and child behavior problems using a sample of intact, single mother, and stepparent families. Although findings are attenuated in this study relative to twin and adoption studies, the authors suggest that genetic influences are important to child behavior problems and partner-child negativity. Shared environmental influences are implicated in mother-child and partner-child negativity.

Preliminary data from an adoption study of 5-10-year-old children is provided by Deater-Deckard, Petrill, and Wilkerson. This study not only examines environmental influences on social and cognitive development using an adoption design, because many of the children are adopted from overseas, but also examines the issue of international adoption.

Manke and Pike take this issue even further, noting that it is also possible to more explicitly use the perceptions of different raters to better understand social relations. The authors conduct an innovative analysis examining the genetic and environmental influences using the Social Relations Model (SRM: Kenny & LaVoie, 1984).

Empirical Papers Examining Peer Influences

In her recent review, Harris (1995) suggests that researchers must take into account outside of the home influences if we are to develop a full understanding of the role of social relationships on personality development. The following manuscripts examine how genes and environments impact peer relationships. Three such articles are included in this collection.

The article by Gilson, Hunt, and Rowe examines whether delinquency and verbal intelligence are influenced by peer selection/influence or social homogamy. The selection/influence models suggest that a child's outcomes are shaped by particular peers while the social homogamy model suggests that more distal aspects of the environment are affecting outcomes. The results of this study suggest that delinquency is best predicted by a selection/influence model while verbal intelligence is best predicted by a social homogamy model.

Edelbrock (1983) and others have suggested that behavior is a function of the context in which it is measured. Thus, another important issue raised by developmental researchers is the need to use multiple methods of assessment. While many of the papers in this collection employ multiple raters or methods, Leve's article examines peer relationships using a variety of methods: Observation, parent-report, self-report, teacher reports, and records.

O'Connor, Jenkins, Hewitt, DeFries, and Plomin examine the relationship between the quality of the parent-child relationship and the quality of child-peer relationship. This relationship was examined through parent- and teacher-report using a longitudinal sample of adoptive children and their families. Interestingly, the authors employ a multilevel model approach to examine developmental change. The results of this study suggest that parental warmth/support and family expressiveness at age 10 predict teacher-rated peer popularity at age 12. Conversely, negative parental control at 10 years of age predicted increased peer problems at 12 years of age.

Integrating Behavioral Genetics and Theory of Social Relationships

In the final selection of this collection, Robert Plomin and Kathryn Asbury discuss these manuscripts in order to draw conclusions concerning how the study of the environment and social relationships is predicated by the use of genetically sensitive designs.

REFERENCES

DeFries, J. C., & Fulker, D. W. (1985). Multiple regression analysis of twin data. *Behavioral Genetics, 15*, 467-473.

DeFries, J. C., & Fulker, D. W. (1988). Etiology of deviant scores versus individual differences. *Acta Genetica, Med Gemellol, Twin Res, 37*, 205-216.

Edelbrock, C. (1983). Problems and issues in using rating scales to assess child personality and psychopathology. *School Psychology Review, 12*, 293-299.

Harris, J. R. (1995). Where is the child's environment? A group socialization theory of development. *Psychological Review, 102*(3), 458-489.

Kenny, D. A., & LaVoie, L. (1984). The social relations model. In L. Berkowitz (Ed.), *Advances in experimental social psychology* (pp. 141-182). Orlando, FL: Academic Press.

Neale, M. C. (1995). *Mx: Statistical modeling* (3rd ed.). Box 710 MCV, Richmond, VA 23298: Department of Psychiatry, Medical College of Virginia.

Neale, M. C., & Cardon, L. R. (1992). *Methodology for genetic studies of twins and families.* Dordrecht: Kluwer Academic Publishers.

Plomin, R., DeFries, J. C., McClearn, G. E., & McGuffin, P. (2000). *Behavioral genetics* (4th ed.). New York: Worth Publishers.

Genetic and Environmental Influences on Parenting and Marital Relationships: Current Findings and Future Directions

Hilary Towers
Erica L. Spotts
Jenae M. Neiderhiser

SUMMARY. There have been an increasing number of studies that have examined genetic influences on measures of family relationships, which have typically been conceptualized as environmental. These measures include parenting, sibling relationships, marital relationships and divorce. The current review discusses these findings and presents newly emerging results that suggest the mechanisms that may be involved, at least in mother-child relationships. The best way to understand how genetic factors can influences "environmental" measures is through genotype-environment correlation. The different types of genotype-environment correlation that are likely to be operating in each type of relationship are described and the implications of these findings are discussed. Finally, new studies that are currently underway are described including a longitudinal study of

---

Hilary Towers, Erica L. Spotts, and Jenae M. Neiderhiser are affiliated with the Center for Family Research, George Washington University, Washington, DC. Hilary Towers' research was supported by NRSA grant number 5F31MHQ12788.

Address correspondence to: Jenae M. Neiderhiser, Center for Family Research, George Washington University, 2300 Eye Street, N.W., Room 613 Ross Hall, Washington, DC 20037.

[Haworth co-indexing entry note]: "Genetic and Environmental Influences on Parenting and Marital Relationships: Current Findings and Future Directions." Towers, Hilary, Erica L. Spotts, and Jenae M. Neiderhiser. Co-published simultaneously in *Marriage & Family Review* (The Haworth Press, Inc.) Vol. 33, No. 1, 2001, pp. 11-29; and: *Gene-Environment Processes in Social Behaviors and Relationships* (ed: Kirby Deater-Deckard, and Stephen A. Petrill) The Haworth Press, Inc., 2001, pp. 11-29. Single or multiple copies of this article are available for a fee from The Haworth Document Delivery Service [1-800-HAWORTH, 9:00 a.m. - 5:00 p.m. (EST). E-mail address: getinfo@haworthpressinc.com].

family relationships that is currently collecting DNA in an effort to better disentangle genotype-environment correlations and interactions. *[Article copies available for a fee from The Haworth Document Delivery Service: 1-800-HAWORTH. E-mail address: <getinfo@haworthpressinc.com> Website: <http://www.HaworthPress.com> © 2001 by The Haworth Press, Inc. All rights reserved.]*

KEYWORDS. Genetics, family relations, parenting, marriage

The study of interpersonal relationships has long been a priority of psychological research, both developmental and clinical in nature. For example, parent-child relationships have been described as both critical (e.g., Bowlby, 1969; Maccoby, 1992; Patterson, DeBaryshe, & Ramsey, 1989) and essentially inconsequential (Harris, 1998) to the adjustment of children and adolescents. Interest in sibling relationships has spawned numerous investigations, including the potential influence of siblings on adjustment (Cicirelli, 1994), the role of siblings as helpers, or models (Dunn, 1996), and the evolving nature of sibling relationships throughout the lifespan (Gold, 1996). Marital relationships have been found to be important correlates of the mental health of each partner, whether the association is characterized in terms of the effect of marriage on individuals' mental health status (Beach & Cassidy, 1991; Seagraves, 1980), or vice-versa (Block, 1981; Dobson, Jacobson & Victor, 1988). Finally, peer relationships have been held as potentially *the* most critical source of influence on the development and adjustment of children and adolescents (Buhrmester & Furman, 1987; Harris, 1998). The focus of the current review is on family relationships, in particular, parent-child and marital relationships. These relationships have been the focus of family research for some time, but until recently have not been examined using genetically sensitive designs. By examining genetic and environmental contributions to relationships among family members, we hope to better understand the mechanisms by which these relationships impact individual development and adjustment.

PARENT-CHILD RELATIONSHIPS

Despite assertions of recent theorists to the contrary (Harris, 1998), both clinical and developmental research has demonstrated the unequivocal impact of parents on the development and adjustment outcome of their children. This influence first becomes manifest in secure or insecure attachment in infancy and early childhood, a parent-child bond now recognized to have significant implications for later formation of relationships with peers (Elicker, Englund, & Sroufe, 1992), social and emotional involvement at school (Sroufe, Fox, & Pancake, 1983), behavior problems (Gomez & Gomez, 2000; Shaw & Vondra, 1995; Patterson et al., 1989) and even academic performance in medical school (Hoeschl & Kozeny, 1997). Later, parents play an important role in shaping the educational and vocational choices their adolescents make (Sebald, 1986), the kinds of value systems they adopt (Brook, Brook, Gordon, Whiteman, & Cohen, 1990), and the sorts of peers with whom they choose to congregate (Fuligni & Eccles, 1993). The majority of studies that have examined associations among parent-child relationships and child and adolescent adjustment have assumed, based on various theoretical positions, that associations between family environments and child behavior are primarily unidirectional in nature: from "environment" to child. This has been particularly true of socialization studies examining parent-child relationships, where associations between parenting styles and children's behaviors have been, until fairly recently, almost uniformly interpreted as evidence of a parent to child effect (Hirschi, 1971; Patterson et al., 1989). For example, proponents of control theory assert that association between a parent's failure to provide supervision or tendency to discipline a child harshly and the child's tendency to act out in school should be interpreted as evidence of the parent's behavior having *caused* the child's behavior (Hirschi, 1971). While such hypotheses are intriguing, and potentially accurate, they are difficult to test given the reliance of traditional parent-child studies upon assessment of only one child per family. A longitudinal study of the association between inter-parental conflict and young adult adjustment, for example, found "direct effects for inter-parental conflict and the parent-young adult relationship on young adult functioning" (Neighbors, Forehand, & Bau, 1997).

A potential problem with assuming that parenting predicts child and adolescent outcome is that most such studies hinge upon assessment of a single target child in the family (rather than on a comparison of two siblings, for instance). Therefore, the ability to detect differential parent-child relationships and potential responses of the parents to each child (which may be due, in part, to genetic influences) is unavailable. This oversight is especially critical given the recent genetic research that has found that both child and *parenting* behaviors are moderately to substantially genetically influenced (Deater-Deckard & O'Connor, 2000; Leve, Winebarger, Fagot, Reid, & Goldsmith, 1998; Neiderhiser, Reiss, Hetherington & Plomin, 1999; Plomin, 1994). While finding genetic influences on child and adolescent behavior may be "old news" to some, finding that measures typically thought of as "environmental" (e.g., parenting) are influenced by genetic factors–emanating from the *child*, no less–came as a surprise to many in the various sub-fields of psychology.

GENETIC AND ENVIRONMENTAL INFLUENCES ON PARENT-CHILD RELATIONSHIPS

One of the first assessments of genetic influences on parenting examined whether or not parents treated MZ and DZ twins differently and, if so, whether the differential treatment occurred as a response to actual genetic differences in the twin types, or as the result of perceptual "biases" on the part of parents (Lytton, 1977). It was determined that parents did treat MZ twins differently than DZ twins, and that the differential treatment seemed to be due to genetic differences (and not just parental perceptions of differences). A more comprehensive analysis of both genetic and environmental influences on parenting was conducted by David Rowe in the early 1980s (1981; 1983). Rowe used two independent samples of adolescent twins to examine the extent to which genetic and environmental factors contributed to adolescent self-reports of their family environment, which included various parenting dimensions encompassed within the broad factors of acceptance-rejection and control. Rowe found evidence of genetic influences for parental warmth and shared environmental influences (nongenetic factors that make family members similar to one another) for parental control. Interestingly, this study also found that environmental factors the twins did *not* share (located either within or outside of the home) accounted for approximately half of the variance in the twins' perceptions of parental warmth and control. These factors, known in the quantitative genetic literature as "nonshared environmental factors," may reflect differential treatment of twins by their parents, different classroom experiences, or different peer relationships and also include error.

The finding of genetic influence on child and adolescent perceptions of parental warmth, and little or no genetic influence on perceptions of parental control, has been replicated in several different reports using a variety of different samples (e.g., Elkins, McGue, & Iacono, 1997; Goodman & Stevenson, 1989; Plomin, McClearn, Pedersen, Nesselroade, & Bergeman, 1989). Particularly convincing is the finding of such a pattern using child (Braungart, 1994), adolescent (Elkins et al., 1997; Goodman & Stevenson, 1989), and adult (Plomin et al., 1989) samples. Findings from a study using retrospective reports of childhood rearing environment from adult twins substantiate the claim that genetic factors do influence parenting behaviors differentially, even when relying on reports of behaviors that occurred many years ago (Plomin et al., 1989). It should be noted that one exception to the finding of primarily genetic influences on measures of parental warmth and primarily shared environmental influences on parental control comes from the Nonshared Environment in Adolescent Development (NEAD; Reiss et al., 1994) project. Several papers from the NEAD project have found substantial genetic influences on *multiple-rater, multiple-measure* composites of parental positivity (warmth), negativity, and monitoring/control (Neiderhiser et al., 1999; Pike, McGuire, Hetherington, Reiss & Plomin, 1996; Reiss et al., 2000). Shared and nonshared environmental influences were also significant and substantial for these composites of parenting in the NEAD sample. It is possible that the use of composites across multiple raters and measures results in a construct that is more consistent across different settings that is also more likely to be influenced by genetic factors. There is some evidence that this may be the case, in that when single-rater constructs are examined genetic influences are much smaller for parental control (Plomin, Reiss, Hetherington, & Howe, 1994; O'Connor, Hetherington, Reiss & Plomin, 1995).

The findings described thus far are from *child-based* designs. In other words, the child's genes are the focus of analysis. While child-based genetic designs have been useful in establishing the importance of genetic factors to parent-child relationships, they are unable to specify the mechanisms by which these genetic factors operate. It is tempting to conclude, based on the literature reviewed up to this point, that it is primarily the influence of the *child's* genes reflected in the finding of genetic influence on parental warmth (since it is the child's genes, after all, which are the focus of the design). The notion that children are active creators of their own destinies is one that makes intuitive sense, and is appealing to many developmentalists and geneticists (e.g., Scarr & McCartney, 1983). It is important, however, to be cautious in drawing such a conclusion based on results of child-based designs alone, since the finding of genetic influence on parenting within these designs does not preclude the possibility that the *parents'* genes, or simply the 50% genetic similarity between parent and child, are contributing just as much–if not more–to this observed genetic influence. The following section will address this point in detail.

GENOTYPE-ENVIRONMENT CORRELATION AS AN EXPLANATION FOR GENETIC INFLUENCES ON PARENTING

To understand genetic influences on measures of parenting, it is critical to understand the role of genotype-environment correlation. The distinctions between genetic influences on parenting in child-based designs and parent-based designs have also been captured nicely in descriptions of passive and nonpassive genotype-environment correlation (GE correlation; Plomin, DeFries, & Loehlin, 1977; Scarr & McCartney, 1983). Briefly, passive GE correlation refers to a correlation between a measure of the environment (e.g., parenting) and a genotype (e.g., that of a child), which is due to the 50% genetic-relatedness of the two. For example, a father may pass on his "novelty-seeking" genes to his sons. In this case, a passive GE correlation could explain genetic influences on parent-child conflict simply because the same "novelty-seeking" genes in the father and the sons increases the likelihood of conflict in their relationship. Passive GE correlations may (or may not) account for the genetic influences on parenting. In other words, parents may display warmth toward their children because of some genetic factor (perhaps personality) that originates within the *parent*. A child-based design is not able to distinguish such a mechanism from that of a nonpassive GE correlation mechanism, because parents who are genetically related to the children also provide the parenting under study (with the exception of adoption designs, described in more detail below).

It is also possible that children's genetically influenced characteristics play an important role in "shaping" their environments, including the behaviors of their parents. Such a mechanism might be reflected in one of two types of nonpassive GE correlation: evocative or active. Evocative GE correlation refers to the reaction of others to an individual based upon some genetically influenced trait of that individual. For example, a mother may react with cold or inconsistent parenting to a temperamentally difficult infant or child. Active GE correlation refers to an individual's seeking out a particular environment, based upon a genetically influenced characteristic. The finding of active selection of peers by children and adolescents is one example of active GE correlation (Manke, McGuire, Reiss, Hetherington, & Plomin, 1995). As evocative and active GE correlations are methodologically indistinguishable, they are often referred to collectively as "nonpassive" GE correlation (Neiderhiser et al., 2001; Spotts & Neiderhiser, in press).

How, then, can passive and nonpassive GE correlation be distinguished from one another? Child-based studies do not allow for such a distinction because the child's genes, only, are the unit of measurement. One way to begin to differentiate between passive and nonpassive GE correlation is through the use of an adoption design, in which the behaviors of the biological parents of an adopted child and those of the adoptive parents are correlated. When such a correlation exists, and the behavior of the child mediates this association, it provides strong evidence that there is something about *the child* that causes the adoptive parents' behavior to correlate with characteristics of the biological parents–implicating a nonpassive genotype-environment correlation. This strategy has been used in two recent reports from two different adoption studies. In both studies, correlations between the biological parents' disorders (a proxy for genotype) and the adoptive parents' parenting style (a proxy for environment) were found (Ge et al., 1996; O'Connor, Deater-Deckard, Fulker, Rutter, & Plomin, 1998). When the adopted child's behavior was entered into the regression equation as a mediator the association between the biological parents' characteristics and the adoptive parents' parenting was significantly reduced, indicating the presence of evocative genotype-environment correlation.

Another report from the Colorado Adoption Project systematically examined Scarr's theory (1992) that GE correlation changes throughout development from mostly passive in infancy to mostly nonpassive in early adolescence (Spotts & Neiderhiser, in press). The findings did not support this developmental theory of change in GE correlation. While some evidence of nonpassive GE correlation was present for child IQ at age 10 and for talent and psychopathology at age 11, passive GE correlation was indicated for externalizing, sociability, and general self-worth at years 10, 11 and 12. Notably, an earlier analysis of the same adoption sample during middle childhood (ages 7 to 9) found passive GE correlations to be important for externalizing and behavior problems, although the same report also showed increasing nonpassive GE correlations throughout middle childhood (Hershberger, 1994). Overall, these studies seem to indicate that the presence or absence of passive and nonpassive GE correlation at a given age operates under a domain-specific, as opposed to general developmental, mechanism. In other words, we might do well to refrain from making broad generalizations about the "overall" pattern, or direction, of genetic and environmental influences in parent-child relationships across time, and instead focus on particular domains of behavior.

Another method that can be used to distinguish passive from nonpassive GE correlation is to employ parent-based twin or sibling designs. Parent-based designs are samples that focus on genetic influences on parents from the parents' genes. For instance, a parent-based design might entail an assessment of parenting behaviors among adult twins toward their adolescent children. Alternatively, the more commonly used child-based designs revolve around a comparison of twins or siblings who are children. The parent-based twin design is especially useful when the measures and sample compliment a child-based design. If genetic influences are found for measures of parenting in a parent-based design, this suggests that there is some genetically-influenced characteristic of the parent that influences their parenting. If shared environmental influences on the *same measures of parenting* are found using a child-based design, then it is likely that the main mode of genetic influence is from parent to child, or through passive GE correlation. If, however, genetic influences on parenting are found in a child-based study, but not in a complimentary parent-based study, this would suggest that the mechanism of genetic influence is running from child to parent, or is nonpassive in nature. Such a finding rules out the possibility that *parent's* genes were responsible for the observed genetic influence.

PARENT-BASED STUDIES OF PARENTING

To date, there have been few parent-based studies of parenting. Among those studies that are parent-based, some interesting patterns of findings have emerged. In particular, it appears that passive GE correlation may be a viable explanation for parenting—at least with respect to early parent-child relationships. A recent study of parenting by adult twin parents toward their eight-year-old children showed evidence of genetic and nonshared environmental influences on both warmth and control measures of parenting (Losoya, Callor, Rowe & Goldsmith, 1997). Such a finding is intriguing and novel in several respects. First, it suggests that parent's genes are contributing substantially to their own behavior as parents. While this does not "undo" the finding of children's genetic influences on parenting in child-based designs, it does call into question the idea that parenting is purely a *response to* children's genetically influenced tendencies.

Second, the finding of genetic influences on parental control in a parent-based study, but not in most child-based studies, sheds some light on the way in which parental control may exert itself in family relationships. It seems that for controlling behavior, parents' genes may be the "driving force," with children's genetically influenced tendencies having little influence on this behavior. Thus, parent-based studies to date suggest that an observed association between parental control and child behavior may be attributable to a passive GE correlation mechanism. In fact, this finding has been replicated in another parent-based study of current global family environment (using the Family Environment Scale, or FES) of older adult twins in Sweden (Plomin et al., 1989). In this analysis, adults' perceptions of their current global family environment, including expressiveness, cultural orientation, organization and *control* showed significant genetic influences. The finding of genetic influences on dimensions of parental control has since been replicated in two independent studies employing parent-based designs (Perusse, Neale, Heath, & Eaves, 1994; Losoya et al., 1997), although a recent analysis of adult twin women did not replicate this finding (Kendler, 1996).

The Twin Mothers project (TM) was designed, in part, to use a parent-based design to examine parenting in a sample of 326 pairs of adult twin women with adolescent children. The measurement and the sampling criterion were designed to mirror the NEAD project as much as feasible. A recent report from this project compared the genetic and environmental influences on mothering as estimated from the parent-based TM sample to the child-based NEAD sample (Neiderhiser et al., 2001). This is the first such study to directly compare child-based and parent-based designs focused on the same set of mother-child relationships. The findings will be summarized using mothers' reports as an example. For mother's warmth, the parent-based TM sample found genetic and nonshared environmental influences only, while the child-based NEAD sample found genetic, shared and nonshared environmental influences to be important. Genetic and nonshared environmental influences on mother's negativity in the TM sample were present, whereas in the child-based NEAD study, findings for negativity were similar to those for warmth in that a substantial proportion of the variance was explained by genetic influences with the remaining variance split nearly equally between shared and nonshared environmental influences. Finally, for control, the parent-based TM study found no evidence of genetic or shared environmental influences while genetic influences were significant in the child-based NEAD sample. Monitoring showed genetic influences for both samples, with the bulk of the variance due to nonshared environmental influences for the TM sample and mostly shared environmental influences for NEAD.

The findings from this comparison study reveal several interesting and novel patterns of influences with regard to parent-child relationships–again, specific to each particular domain of parenting behavior. One surprising finding was a *lack* of substantial genetic influences on measures of control in the parent-based TM sample, suggesting that passive GE correlation is not a viable explanation for genetic influences on mother's control of her adolescent. This finding, while at odds with previous parent-based studies of parenting (Losoya et al., 1997; Perusse et al., 1994), is consistent with another parent-based study of adult twin women (Kendler, 1996), which found genetic influences on measures of parental warmth, but not on measures of control. One possible explanation for this discrepancy discussed by Neiderhiser and colleagues is that the findings are specific to *mother-adolescent* relationships since they replicate findings of studies that examined mother-adolescent dyads (Kendler, 1996), but not studies of younger children (Losoya et al., 1997), or both father- and mother-child dyads together (Perusse et al., 1994).

Contrary to the findings for control, genetic influences on maternal positivity and negativity were approximately the same across the two samples, suggesting that passive GE mechanisms may be important for mother's positivity and negativity towards her adolescent. These two findings, taken together, suggest that different parenting constructs are based upon different mechanisms. For parental control, at least with regard to mothering of adolescents, it seems that adolescents' genetically-influenced traits and behaviors can and do elicit certain control-related parenting behaviors. This does not seem to be true for parental positivity (e.g., warmth), which appears to be much more under the genetic "control" of mothers of adolescents although there is also evidence of evocative GE correlation given the sizable genetic influence for the NEAD sample.

Finally, perhaps the most intriguing finding reported by Neiderhiser et al. (2001) was the sizable amount of nonshared environmental influences found for mothering in the TM sample, especially when compared to the more modest levels of nonshared environmental influences found in the NEAD sample. This finding makes sense intuitively, since adult twin women have had ample opportunity and time to gain independence from one another, in terms of creating their own "niches" in life–different friends, different careers, different spouses, and different children of their own. All of these differences are likely to constitute sources of nonshared environmental influences on parenting behaviors. Future research is needed to isolate and identify these specific sources of nonshared environmental influences on parenting. With regard to the nature of parent-child relationships, on a broader level of analysis, our findings to date concerning the direction of genetic effects on parent-child associations remain somewhat inconclusive. To begin to piece together the particular workings of specific associations across varying age groups may be a reasonable starting point for clarifying the "big picture."

GENETIC AND ENVIRONMENTAL INFLUENCES ON MARITAL RELATIONSHIPS

One way to begin to uncover sources of influence on parenting behavior is to examine the genetic and environmental contributions to specific relationships likely to be associated with parenting. Perhaps the most obvious of these is the marital relationship, which has repeatedly been documented as being intertwined with parent-child relationships (Emery & Tuer, 1993; Dunn, Deater-Deckard, Pickering, Golding, & ALSPAC Study Team, 1999). Marital quality and mental health, particularly depression, have long been found to be associated (e.g., Sweatman, 1999; Sandburg & Harper, 1999). There is also evidence that women are more affected by this association than men (Horowitz, McLaughlin & White, 1998). Supportive relationships can buffer clinical depression (Cutrona, 1996), though problematic relationships seem to have more of an effect, albeit a negative one, than supportive ones (Horowitz et al., 1998). The direction of effects for these associations is not entirely clear, however. Dehle and Weiss (1998) found that initially low marital quality predicted an increase in depressed mood three months later. Conversely, the same study found that initially higher scores of depressed mood predicted greater declines in marital quality during the same time period. Poor marital quality is also associated with phobias, panic disorder, and generalized anxiety disorder, though interestingly, spouses concordant for phobias have higher marital quality (McLeod, 1994).

While research assessing general trends in marital quality has been plentiful, studies examining individual differences in, or the mechanisms underlying, marital quality remain scarce (Karney & Bradbury, 1995). To date, there have been two studies examining genetic and environmental influences on *divorce* (Jocklin, McGue, & Lykken, 1996;

McGue & Lykken, 1992). Notably, these studies found evidence for genetic influences on risk for divorce (McGue & Lykken, 1992), with personality characteristics accounting for a sizeable portion of the genetic influence on divorce (Jocklin et al., 1996). The following provides a brief summary of the major findings of the first study ever designed to assess mechanisms underlying individual differences in *marital quality* (Spotts et al., 2001).

One of the aims of the Twin Mothers project, described above, was to examine how marital satisfaction impacted maternal mental health. Toward this aim, the Dyadic Adjustment Scale (DAS; Spanier, 1976) and the Marital Adjustment Test (MAT; Locke and Wallace, 1987), both often-used measures of marital quality, were administered to both the twin women and their spouses.[1] The analyses summarized here examined both the genetic and environmental influences on wives' report of marital quality and husbands' report of marital quality. Because the wives were the twins, the only genetic and environmental influences that can be described are those of the wives. This means that we can explore how wife-based genetic and environmental influences affect both her perceptions of marital quality and how her husband's perceptions reflect her genetically and environmentally influenced characteristics.

Spotts and colleagues (2001) found that most measures of wives' marital quality were at least moderately influenced by genetic factors (standardized parameter estimates of .24-.33), and primarily influenced by nonshared environmental influences (standardized parameter estimates of .67-.85). One exception was DAS Affectional Expression, which showed negligible amounts of genetic influence. Shared environmental influences played no role in the twin women's reports of their marital quality. Results were similar for husband's reports of marital quality. Again, any influences on husband reports were those of the *wives'* genetically and environmentally influenced characteristics. Genetic influences were found for all but DAS Affectional Expression and MAT Agreement on Parenting. Nonshared environmental influences were found for all measures of marital quality, and shared environmental influences were not found.

This report took the analyses one step further by asking whether or not the genetic and environmental factors that influence the wife's report of marital quality are the same factors that influence the husband's report of marital quality. All of the genetic influences on both wife and husband reports of Affectional Expression and Dyadic Satisfaction overlapped. One way to interpret this finding is that the same genetically influenced characteristics of the women influenced their own reports of the affectional expression and dyadic satisfaction as well as their husbands' reports of the same marital quality characteristics. For the remaining scales, the same genetic factors accounted for some of the overlap between wife and husband, but some of the wife's report was influenced by genetic factors independent of those that influenced husband report. This was also the case for nonshared environmental factors; the same nonshared environmental factors influenced, in part, both wife and husband reports of marital quality, but some of the wife's report was influenced by factors that did not also influence husband reports. This suggests that there are general genetic and nonshared environmental factors that are equally important for both husband and wife reports of marital quality. These factors could originate in the wife, which would be an example of GE correlation. These factors could also be the result of assortative mating, in that the husband and wife selected each other for their phenotypic characteristics. There are also, of course, member-specific factors that influence perceptions of marriage; some characteristics of the wife influence her marital quality without being reflected in the husband's reports of marital quality. It is likely that a husband-based twin design would find a similar pattern of results in regard to member-specific influences on marriage vs. those that are shared for both members of the couple.

CONCLUSIONS AND FUTURE STUDIES

Taken together there are two broad conclusions that can be drawn from studies of family relationships that have employed genetically sensitive designs. First, simply calling a measure "environmental" or measuring an aspect of the family does not mean that genetic factors do not play a role. The majority of studies that have examined genetic influences on measures of family environment have found evidence for substantial and significant genetic, as well as environmental, influences on these measures. There is a great deal more to be learned about how family relationships are influenced by genetic factors. For example, some researchers have suggested that the family, through their relationships and interactions with one another, may actually help to determine which genetically influenced characteristics are expressed (Reiss, Neiderhiser, Hetherington & Plomin, 2000). This possibility will be discussed in more detail below. It is important to also acknowledge the substantial proportion of variance in family relationships that is *not* due to genetic factors. In parent-child relationships this environmental variance is usually split nearly evenly between shared and nonshared environmental influences, with some variation depending upon whose report is examined and the construct under study. The one exception to this pattern of findings comes from the few parent-based studies of parent-child relationships that have been reported. In these reports, the largest component of environmental variance is nonshared environmental factors. For marital relationships, although one might expect there to be some evidence of shared environmental influences due, perhaps, to modeling of the current relationship on their parents relationships, there is evidence only of nonshared environmental influences. In fact, nonshared environmental influences explain the bulk of the variance in marital relationship quality and satisfaction.

The second broad conclusion has to do with the type of genotype-environment correlation that is likely to be operating. The combination of the child-based NEAD sample with the complementary parent-based TM sample allows primarily evocative genotype-environment correlations to be implicated in the majority of mother-child relationships. The one exception to this is for maternal warmth, which does not permit a clear conclusion in regard to passive or evocative G-E correlation mechanisms. Finding genetic influences on husband reports of marital quality in the TM project also suggests the importance of evocative G-E correlations in marital interactions, though these results could also be explained in terms of assortative mating.

Currently, a study is underway that may provide additional specification of the mechanisms involved in the mediation and moderation of gene expression through family relationships. This study is reassessing the NEAD sample and collecting DNA in order to examine candidate genes that are likely to be associated with many of the behavioral measures that have already been collected in this sample (Neiderhiser, 1999). The strategy of examining candidate genes in family samples has been described in detail elsewhere (e.g., Plomin & Rutter, 1998). In brief, such a strategy entails associating selected genes with specific behaviors, through techniques as basic as regression analyses. If an association is found between a particular candidate gene and a behavior then that is evidence that that gene is important in influencing that particular behavior. More relevant to the current report are the additional analyses that enable the specification of the possible role of family relationships in mediating (genotype-environment correlation) or moderating (genotype-environment interaction) the expression of these candidate genes for a particular behavior. Because both the family relationship (environment) and the candidate gene are directly measured, it is possible to detect such interactions or correlations where they had been undetectable in traditional "anonymous" designs (e.g., twin and sibling designs).

Until studies like that described above are completed and replicated, we must rely on a combination of techniques and designs to understand the mechanisms that are involved in explaining genetic influences on measures of family relationships. To date such studies suggest that not only are genetic factors important in both parent-child and marital relationships, but that the processes involved differ based on which aspects of the relationship are examined (e.g., warmth and negativity vs. monitoring). As more longitudinal studies are conducted that include both genetically sensitive designs and careful measurement of family relationships, these mechanisms will come into sharper focus.

It is important to note that many developmental researchers who have typically considered themselves to be "environmentalists" are beginning to take note of the findings from behavioral genetics. For example, two recent reviews of theory and research in parenting stressed the importance of considering the role of children's genetically influenced characteristics when examining the role of parenting while also cautioning researchers in behavioral genetics to carefully and directly measure the environment and to consider the two together whenever possible (Collins, Maccoby, Steinberg, Hetherington, & Bornstein, 2000; Maccoby, 2000). This attention and acknowledgement is welcome and critical in increasing our understanding of the role of both children and parents in shaping development. There is no similar acknowledgement from family researchers of the importance of genetic factors in influencing marital relationships, but such reports are only just beginning to emerge. It is clear that genetic and environmental factors are important in shaping the development of individuals. The critical question is how these factors operate together.

NOTE

1. Note that 6% of the sample were cohabiting without being married; cohabitation is a common and accepted alternative to marriage in Sweden. Both the DAS and the MAT have been validated on married and unmarried couples.

REFERENCES

Beach, S. R., & Cassidy, J. F. (1991). The marital discord model of depression. *Comprehensive Mental Health Care, 1*, 119-136.

Block, A. R. (1981). An investigation of the response of the spouse to chronic pain behavior. *Psychosomatic Medicine, 43*, 415-422.

Bowlby, J. (1969). Disruption of affectional bonds and its effects on behavior. *Canada's Mental Health Supplement, 59,* 12.

Braungart, J. M. (1994). Genetic influences on "environmental" measures. In J. C. DeFries, R. Plomin, & D. W. Fulker (Eds.), *Nature and nurture during middle childhood* (pp. 233-248). Cambridge, MA: Blackwell.

Brook, J. S., Brook, D. W., Gordon, A. S., Whiteman, M., & Cohen, D. (1990). The psychosocial etiology of adolescent drug use: A family interactional approach. *Genetic, Social, & General Psychology Monographs, 116,* 111-267.

Buhrmester, D. & Furman, W. (1987). The development of companionship and intimacy. *Child Development, 58,* 1101-1113.

Cicirelli, V. G. (1994). The longest bond: The sibling life cycle. L'Abate Luciano (Ed). *Handbook of developmental family psychology and psychopathology. Wiley series on personality processes* (pp. 44-59). New York, NY: John Wiley & Sons.

Collins, W. A., Maccoby, E. E., Steinberg, L., Hetherington, E. M., & Bornstein, M. H. (2000). Contemporary research on parenting: The case for nature and nurture. *American Psychologist, 55,* 218-232.

Cutrona, C. E. (1996). Social support as a determinant of marital quality: The interplay of negative and supportive behaviors. In G. R. Pierce, B. R. Sarason & I. G. Sarason (Eds.), *Handbook of social support and the family. Plenum series on stress and coping* (pp. 173-194). New York, NY: Plenum Press.

Deater-Deckard, K. & O'Connor, T. G. (2000). Parent-child mutuality in early childhood: Two behavioral genetic studies. *Developmental Psychology, 36,* 561-570.

Dehle, C. & Weiss, R. L. (1998). Sex differences in prospective associations between marital quality and depressed mood. *Journal of Marriage and the Family, 60,* 1002-1011.

Dobson, K. S., Jacobson, N. S., & Victor, J. (1988). Integration of cognitive therapy and behavioral marital therapy. In J. F. Clarkin, G. L. Haas, & I. D. Glick (Eds.), *Affective disorders and the family: Assessment and treatment* (pp. 53-88). New York: Guilford Press.

Dunn, J. (1996). Siblings: The first society. In N. Vanzetti & S. Duck (Eds.), *A lifetime of relationships* (pp. 105-124). Pacific Grove, CA: Brooks/Cole Publishing Co.

Dunn, J., Deater-Deckard, K., Pickering, K., Golding, J., & ALSPAC Study Team (1999). Siblings, parents, and partners: Family relationships within a longitudinal community study. *Journal of Child Psychology & Psychiatry & Allied Disciplines, 40,* 1025-1037.

Elicker, J., Englund, M., & Sroufe, L. A. (1992). Predicting peer competence and peer relationships in childhood from early parent-child relationships. In R.D. Parke & G.W. Ladd (Eds.), *Family-peer relationships: Modes of linkage* (pp. 77-106). Hillsdale, NJ: Lawrence Erlbaum Associates.

Elkins, I. J., McGue, M., & Iacono, W. G. (1997). Genetic and environmental influences on parent-son relationships: Evidence for increasing genetic influence during adolescence. *Developmental Psychology, 33,* 351-353.

Emery, R. E., & Tuer, M. (1993). Parenting and the marital relationship. In T. Luster & L. Okagaki (Eds.), *Parenting: An ecological perspective* (pp. 121-148). New Jersey: Lawrence Erlbaum Associates.

Fuligni, A. J. & Eccles, J. S. (1993). Perceived parent-child relationships and early adolescents' orientation toward peers. *Developmental Psychology, 29,* 622-632.

Ge, X., Conger, R. D., Cadoret, R. J., Neiderhiser, J. M., Yates, W., & Troughton, E. (1996). Developmental interface between nature and nurture: A mutual influence model of child antisocial behavior and parenting. *Developmental Psychology, 32,* 574-589.

Gold, D. (1996). Continuities and discontinuities in siblings relationships across the life span. In V. L. Bengston (Ed.), *Adulthood and aging: Research on continuities and discontinuities* (pp. 228-243). New York, NY: Springer Publishing Co.

Gomez, R., & Gomez, A. (2000). Perceived maternal control and support as predictors of hostile-biased attribution of intent and response selection in aggressive boys. *Aggressive Behavior, 26,* 155-168.

Goodman, R., & Stevenson, J. (1989). A twin study of hyperactivity–II. The aetiological role of genes, family relationships and perinatal adversity. *Journal of Child Psychology and Psychiatry, 30,* 691-709.

Harris, J. R. (1998). *The nurture assumption: Why children turn out the way they do.* New York, NY: The Free Press.

Hershberger, S. L. (1994). Genotype-environment interaction and correlation. In J. C. DeFries, R. Plomin, & D. W. Fulker (Eds.), *Nature and nurture during middle childhood* (pp. 281-294). Cambridge, MA: Blackwell.

Hirschi, T. (1971). *Causes of delinquency.* Berkeley, CA: University of California Press.

Hoeschl, C. & Kozeny, J. (1997). Predicting academic performance of medical students: The first three years. *American Journal of Psychiatry, 154*(Supplement), 87-92.

Horowitz, A. V., McLaughlin, J., & White, H. R. (1998). How the negative and positive aspects of partner relationships affect the mental health of young married people. *Journal of Health and Social Behavior, 39,* 124-136.

Jocklin, V., McGue, M., & Lykken, D. T. (1996). Personality and divorce: A genetic analysis. *Journal of Personality and Social Psychology, 71,* 288-299.

Karney, B. R., & Bradbury, T. N. (1995). The longitudinal course of marital quality and stability: A review of theory, method, and research. *Psychological Bulletin, 118,* 3-34.

Kendler, K. S. (1996). Parenting: A genetic-epidemiologic perspective. *American Journal of Psychiatry, 153,* 11-20.

Leve, L. D., Winebarger, A. A., Fagot, B. I., Reid, J. B., & Goldsmith, H. H. (1998). Environmental and genetic variance in children's observed and reported maladaptive behavior. *Child Development, 69,* 1286-1298.

Locke, H. & Wallace, K. (1987). Marital Adjustment Test. In N. Fredman & R. Sherman (Eds.), *Handbook of measurements for marriage and family therapy* (pp. 46-50). New York: Brunner/Mazel, Inc.

Losoya, S. H., Callor, S., Rowe, D. C., & Goldsmith, H. H. (1997). Origins of familial similarity in parenting: A study of twins and adoptive siblings. *Developmental Psychology, 33,* 1012-1023.

Lytton, H. (1977). Do parents create, or respond to, differences in twins? *Developmental Psychology, 13,* 456-459.

Maccoby, E. E. (2000). Parenting and its effects on children: On reading and misreading behavior genetics. *Annual Review of Psychology, 51,* 1-27.

Maccoby, E. E. (1992). The role of parents in the socialization of children: A historical review. *Developmental Psychology, 28,* 1006-1017.

Manke, B., McGuire, S., Reiss, D., Hetherington, E. M., & Plomin, R. (1995). Genetic contributions to adolescents' extrafamilial social interactions: Teachers, best friends, and peers. *Social Development, 4,* 238-256.

McGue, M. & Lykken, D. T. (1992). Genetic influence on risk of divorce. *Psychological Science, 3,* 368-373.

McLeod, J. D. (1994). Anxiety disorders and marital quality. *Journal of Abnormal Psychology, 103,* 767-776.

Neiderhiser, J. M. (1999). Genes, adolescent adjustment, and family processes. NIMH research grant, July 1999-June, 2002.

Neiderhiser, J. M., Reiss, D., Hetherington, E. M., & Plomin, R. (1999). Relationships between parenting and adolescent adjustment over time: Genetic and environmental contributions. *Developmental Psychology, 35,* 680-692.

Neiderhiser, J. M., Reiss, D., Pedersen, N. L., Lichtenstein, P., Hansson, K., Cederblad, M., & Elthammer, O. (2001). Genetic and environmental influences on mothering of adolescents: A comparison of two samples. Submitted for publication.

Neighbors, B. D., Forehand, R., & Bau, J. J. (1997). Interparental conflict and relations with parents as predictors of young adult functioning. *Development & Psychopathology, 9,* 169-187.

O'Connor, T. G., Deater-Deckard, K., Fulker, D., Rutter, M., & Plomin, R. (1998). Genotype-environment correlations in late childhood and early adolescence: Antisocial behavioral problems and coercive parenting. *Developmental Psychology, 34,* 970-981.

O'Connor, T. G., Hetherington, E. M., Reiss, D., & Plomin, R. (1995). A twin-sibling study of observed parent-adolescent interactions. *Child Development, 66,* 812-829.

Patterson, G. R., DeBaryshe, B. D. & Ramsey, E. (1989). A developmental perspective on antisocial behavior. *American Psychologist, 44,* 329-335.

Perusse, D., Neale, M. C., Heath, A. C., & Eaves, L. J. (1994). Human parental behavior: Evidence for genetic influence and potential implication for gene-culture transmission. *Behavior Genetics, 24,* 327-335.

Pike, A., McGuire, S., Hetherington, E. M., Reiss, D., & Plomin, R. (1996). Family environment and adolescent depression and antisocial behavior: A multivariate genetic analysis. *Developmental Psychology, 32,* 590-603.

Plomin, R. (1994). *Genetics and experience: The interplay between nature and nurture.* Thousand Oaks, CA: Sage Publications.

Plomin, R., DeFries, J. C., & Loehlin, J. C. (1977). Genotype-environment interaction and correlation in the analysis of human behavior. *Psychological Bulletin, 84,* 309-322.

Plomin, R., McClearn, G. E., Pedersen, N. L., Nesselroade, J. R., & Bergeman, C. S. (1989). Genetic influence on adults' ratings of their current family environment. *Journal of Marriage and the Family, 51,* 791-803.

Plomin, R., Reiss, D., Hetherington, E. M., & Howe, G. W. (1994). Nature and nurture: Genetic contributions to measures of the family environment. *Developmental Psychology, 30*, 32-43.

Plomin, R., & Rutter, M. (1998). Child development and molecular genetics: What do we do with genes once they are found? *Child Development, 69*, 1225-1242.

Reiss, D., Neiderhiser, J.M., Hetherington, E.M., & Plomin, R. (2000). *The relationship code: Deciphering genetic and social influences on adolescent development.* Cambridge, MA: Harvard University Press.

Reiss, D., Plomin, R., Hetherington, E. M., Howe, G. W., Rovine, M., Tryon, A., & Hagan, M. S. (1994). The separate worlds of teenage siblings: An introduction to the study of the Nonshared Environment and Adolescent Development. In E. M. Hetherington, D. Reiss, & R. Plomin (Eds.), *Separate social worlds of siblings: The impact of nonshared environment on development* (pp. 63-109). Hillsdale, NJ: Lawrence Erlbaum Associates.

Rowe, D. C. (1981). Environmental and genetic influences on dimensions of perceived parenting: A twin study. *Developmental Psychology, 17*, 203-208.

Rowe, D. C. (1983). A biometrical analysis of perceptions of family environment: A study of twins and singleton sibling kinships. *Child Development, 54*, 416-423.

Sandburg, J. G., & Harper, J. M. (1999). Depression in mature marriages: Impact and implications for marital therapy. *Journal of Marital and Family Therapy, 25*, 393-406.

Scarr, S. (1992). Developmental theories for the 1990's: Development and individual differences. *Child Development, 63*, 1-19.

Scarr, S. & McCartney, K. (1983). How people make their own environments: A theory of genotype → environment effects. *Child Development, 54*, 424-435.

Seagraves, R. T. (1980). Marriage and mental health. *Journal of Sex and Marital Therapy, 6*, 187-198.

Sebald, H. (1986). Adolescents' shifting orientation toward parents and peers: A curvilinear trend over recent decades. *Journal of Marriage & the Family, 48*, 5-13.

Shaw, D. S. & Vondra, J. I. (1995). Infant attachment security and maternal predictors of early behavior problems: A longitudinal study of low-income families. *Journal of Abnormal Child Psychology, 23*, 335-357.

Spanier, G. B. (1976). Measuring dyadic adjustment: New scales for assessing quality of marriage and similar dyads. *Journal of Marriage and the Family, 38*, 15-28.

Spotts, E. & Neiderhiser, J. M. (in press). The developmental trajectory of genotype-environment correlation: The increasing role of the individual in selecting environments. In J. Hewitt, R. Plomin & J.C. DeFries (Eds.), *The transition to early adolescence: Nature and nurture.*

Spotts, E. L., Neiderhiser, J. M., Hansson, K., Lichtenstein, P., Cederblad, M., Pedersen, N., Elthammer, O., & Reiss, D. (2001). Genetic and environmental influences on marital relationships. Submitted for publication.

Sroufe, L. A., Fox, N. E., & Pancake, V. R. (1983). Attachment and dependency in developmental perspective. *Child Development, 54*, 1615-1627.

Sweatman, S. M. (1999). Marital satisfaction, cross-cultural adjustment stress, and the psychological sequelae. *Journal of Psychology and Theology, 27*, 154-162.

Nonshared Environment Research: What Is It and Where Is It Going?

Shirley McGuire

SUMMARY. This review focuses on conceptualizations of nonshared environment and on four areas of research that should be targeted for future growth. It is argued that there are at least two different approaches to the study of nonshared environment. "Experience-oriented" researchers center on sibling differential experiences in the family and their role in children's development. "Outcome-oriented" investigators focus on the search for environmental origins of individual differences in outcomes. Turkheimer and Waldron's (2000) concept of objective versus effective nonshared environment and Reiss and colleagues' (2000) notion of single-system versus multi-system nonshared environment processes are also discussed. Four topics for future research are outlined: (1) age-related changes and development; (2) the role of the self; (3) the role of context; and (4) the importance of extrafamilial experiences. More work in these areas will lead to useful theories of how nonshared environment processes are linked to sibling and individual differences in behavioral development and adjustment. *[Article copies available for a fee from The Haworth Document Delivery Service: 1-800-HAWORTH. E-mail address: <getinfo@haworthpressinc.com> Website: <http://www.HaworthPress.com> © 2001 by The Haworth Press, Inc. All rights reserved.]*

Shirley McGuire is affiliated with the Department of Psychology, University of San Francisco, San Francisco, CA 94117-1080.

The author would like to thank Laurie Cartlidge and the reviewers for their helpful comments.

[Haworth co-indexing entry note]: "Nonshared Environment Research: What Is It and Where Is It Going?" McGuire, Shirley. Co-published simultaneously in *Marriage & Family Review* (The Haworth Press, Inc.) Vol. 33, No. 1, 2001, pp. 31-56; and: *Gene-Environment Processes in Social Behaviors and Relationships* (ed: Kirby Deater-Deckard, and Stephen A. Petrill) The Haworth Press, Inc., 2001, pp. 31-56. Single or multiple copies of this article are available for a fee from The Haworth Document Delivery Service [1-800-HAWORTH, 9:00 a.m. - 5:00 p.m. (EST). E-mail address: getinfo@haworthpressinc.com].

KEYWORDS. Nonshared environment, sibling differences, family relationships, parental differential treatment

There was considerable interest in the topic of nonshared environment in the early 1990s with the publication of a book on the subject (Dunn & Plomin, 1990) and an article criticizing the approach (Hoffman, 1991). Nonshared environment is still receiving the attention of family researchers and developmentalists at the beginning of the 21st century. Another book has been published reviewing the results of the largest study of nonshared environment to date (Reiss, Neiderhiser, Hetherington, & Plomin, 2000) and a detailed critique of the area has appeared (Turkheimer & Waldron, 2000). In their review, Turkheimer and Waldron focused on empirical work conducted over the last two decades; many of their conclusions are included in this paper. The present review, however, will focus on different conceptualizations of nonshared environment; that is, what is nonshared environment? It will also describe research areas that are gaining attention and should be targeted for future growth; that is, where is the nonshared environment field going? Specifically, two different definitions of nonshared environment and their associated theoretical issues will be examined and four topics for future research will be outlined: (1) age-related changes and development; (2) the role of the self; (3) the role of context; and (4) the importance of extrafamilial experiences.

WHAT IS NONSHARED ENVIRONMENT?

The term nonshared environment originated in the behavioral genetic literature (see Dunn & Plomin, 1990; Plomin & Daniels, 1987). According to behavioral genetic theory, individual differences in a behavior in a population, such as memory scores, are the result of a combination of inherited genetic differences, shared family experiences, and nonshared environmental experiences (see Plomin, DeFries, McClearn, & Rutter, 1997 for details about behavioral genetic methods). In this research tradition, scientists compare patterns of family resemblance in a behavior across dyads of different degrees of genetic relatedness. They use parent-child, twin-sibling, and adoption designs as a way to understanding the etiology of individual differences in a behavior in a population. Family members may be similar in a behavior because of a shared genetic heritage or because they share experiences in the family (i.e., shared environment), or a combination of both. Nonshared environment is the degree to which family members differ in a trait or behavior. Family members may differ for many reasons: error of measurement, different experiences within the family, unique life events, and inherited genetic differences (Rowe & Plomin, 1981). Nonshared environment is an "anonymous" component of variance in many behavioral genetic studies, however, because the environment is not measured directly. Instead, it is inferred that differential experiences are important because family members differ to a large degree in behavior and traits, even when they are genetically related and live in the same home. In fact, studies of monozygotic (MZ) twins reared in the same family are considered to be the best evidence for the importance of nonshared environment (Pike, Reiss, Hetherington, & Plomin, 1996). MZ twins inherit identical genotypes, and yet MZ twin pairs differ significantly across many important developmental outcomes (e.g., temperament, self-esteem, and cognitive abilities).

One of the strengths of behavioral genetic designs is that variance in a behavior due to shared family influences can be disentangled from variance due to experiences unique to each individual in the family. This can also be done, however, using data from any study that includes multiple family members; that is, data collected using a within-family design. Attitudes or behaviors are compared within the family (e.g., parents versus their children, older sibling versus younger sibling, twin A versus twin B) as well as across different families (e.g., children with controlling parents versus children with warm parents). In many socialization studies, only one child of each family is included in the study and their behavior is compared across different families. This is called a between-family design. For example, a researcher using a between-family design would examine the association between children's exploratory play and parents' child-rearing behaviors. All the children would come from different families. A researcher using a within-family design would investigate the association between the two siblings' exploratory behavior styles. Behavioral geneticists would take the within-family design one step further and compare sibling similarity in exploratory behavior across pairs of differ degrees of genetic relatedness (e.g., monozygotic twins, dizygotic twins, full siblings, half siblings, unrelated pairs). Within-family designs expand the amount of information that can be learned about links between family experiences and children's development.

In the broadest sense, nonshared environment is the study of within-family differences (e.g., different experiences, different outcomes, different roles, and different perceptions across family members). Several research traditions have painted the family as a collection of interrelated individuals with distinct viewpoints and experiences. Some of this work predates behavioral genetic research studies documenting sibling differences in behavior. For instance, marriage and family researchers have shown that two people can have very different perceptions of a "shared" relationship (e.g., Bernard, 1972). Family systems theorists have often discussed the relevance of unique family roles for understanding individual functioning (e.g., Bowen, 1978). Psychoanalytic theories have provided a place for sibling rivalry and deidentification in personality development (e.g., Schachter, Gilutz, Shore, & Adler, 1978). In this sense, the family has been thought of as a nonshared context by some since the beginning of modern psychology and family research. Still, it is fair to say that the field did not really gain momentum until Plomin and Daniels' (1987) review of behavioral genetic studies using sibling designs. They showed that siblings were not similar in important developmental outcomes and adjustment, even after controlling for genetic factors. Today, most people associate the term "nonshared environment" with sibling differences. There are, however, at least two different approaches to the study of nonshared environment.

TWO APPROACHES TO THE STUDY
OF NONSHARED ENVIRONMENT

Table 1 outlines two perspectives and other conceptual distinctions that have been made in the nonshared environment literature. Some investigators examine within-family differences from a more "experience-oriented" perspective. Researchers using this perspective are interested in children's experiences in the family in relation to their siblings and the correlates and consequences of those experiences. The focus is on *sibling differential experiences in the family* and their role in children's development. Others define nonshared environment from an "outcome-oriented" perspective. They are interested in searching for *the environmental origins of individual differences in outcomes* and use a sibling model to pursue this goal. The experience-oriented and outcome-oriented perspectives are based on two distinct theoretical traditions and have different goals. The approaches, however, are not completely incompatible and some researchers are arguing that aspects

TABLE 1. Definitions of nonshared environment concepts.

Concept	Definition
Approaches:	
Experience-oriented	Primarily concerned with documenting the links between siblings' nonshared experiences inside the family and individual differences in development.
Outcome-oriented	Primarily concerned with understanding nonshared, nongenetic sources of nongenetic sibling differences in outcomes.
Types of Nonshared Environment:[a]	
Objective	Observable differences in siblings' environmental experiences that may or may not result in sibling differences in outcome.
Effective	Environmental experiences that produce sibling differences in outcomes that may be shared or nonshared.
Processes:[b]	
Single-system	Differences in sibling environments (actual or perceived) that are linked to sibling differences in outcome and do not involve social comparison processes.
Multi-system	Differences in sibling environments linked to sibling differences in outcome through the social comparison process.

Note: [a]From Turkheimer & Waldron (2000); [b]From Reiss and colleagues (2000).

of both need to be incorporated into studies of nonshared environment (e.g., Feinberg & Hetherington, 2000; Reiss et al., 2000).

Experience-Oriented Approach

Experience-oriented nonshared environment researchers tend to start conceptually with family experiences and then work forward to understand the implications of these experiences for each child's development. Figure 1 shows a hypothetical distribution of scores on an outcome variable and the types of within-family processes that experience-oriented researchers would investigate (e.g., sibling rivalry, social comparison, and differential roles in the family). The key question is: How do these nonshared environment experiences influence individual development? For example, some investigators have shown that parents treat sons and daughters differently (e.g., Siegel, 1987; White & Brinkerhoff, 1981).

FIGURE 1. Experience-oriented nonshared environment researchers examine sibling differences in family experiences and their relevance for individual development. C = child's score on an outcome measure.

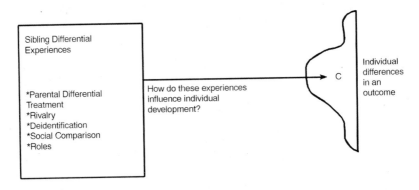

Nonshared environment researchers take this a step further and compare the differential experiences of brothers and sisters in the same family and then link these experiences to individual differences in gender development and adjustment (e.g., Crouter, Manke, & McHale, 1995). This perspective can trace its roots back to constructivist and ecological theories of individual development, such as family systems perspectives, psychoanalytic theories, and person-process-context models (e.g., Bronfenbrenner & Crouter, 1983; Reiss & Oliveri, 1983). The main criticism of this approach is that the studies do not take into account genetic differences between the siblings (Turkheimer & Waldron, 2000). For instance, studies have shown that children who are disciplined more than their siblings have more behavior problems compared to children who are treated equally or better than their siblings (e.g., Conger & Conger, 1994; Dunn, Stocker, & Plomin, 1990; McHale, Crouter, McGuire, & Updegraff, 1995). Parental differential discipline may be a reaction to the siblings' different temperaments. So, one child is being disciplined more than his or her sibling and has behavioral problems because he or she has a difficult temperament. It cannot be assumed that sibling differential experiences are purely a reflection of environmental differences. Some experience-oriented nonshared environment researchers, however, are not concerned about whether these unique experiences are due to genetic or environmental influences. They argue that sibling differences are the product of both factors that cannot be easily disentangled. Instead, the primary issue is the relevance of social processes within the family (e.g., social comparison, sibling rivalry, or sibling deidentification) for individual development. In a sense, it may be more appropriate to call this work the examination of nonshared *experiences* rather than the examination of nonshared environment. This would be consistent with Gottlieb's (e.g., Gottlieb, Wahlsten, & Lickliter, 1998) distinction between the word experience, which means functional activities that are a product of connections across biological and contextual levels, and the word environment, defined as forces that are completely independent of the person. It is also important to realize that these investigators do not claim that sibling differences in experiences will lead to sibling differences in outcomes. In fact, parents may be trying to create just the opposite situation. For example, a father may feel the need to discipline one child more than his or her sibling in order to get them to the same level of social responsibility.

An example of a study using the experience-oriented approach would be McHale and Pawletko's (1992) study of parental differential treatment (i.e., the extent to which a parent treats two siblings in the family differently). They investigated the amount of time children and their siblings spent with their mothers in several activities. Their measure of nonshared environment was a relative difference score created by subtracting the younger sibling's score from the older sibling's score (the difference score is a common measure of parental differential treatment; parental differential treatment is a common measure of nonshared environment). They found different levels of parental differential treatment in families where the younger sibling had a disability compared to families that contained two children with no known handicap. In addition, the links between parental differential treatment and children's adjustment and sibling relationships varied across the two family situations. The focus of the study was to document process-context interactions in relation to parental differential treatment and not to find environmental sources of individual differences in a specific behavior.

Outcome-Oriented Approach

Outcome-oriented nonshared environment researchers, on the other hand, are interested in finding nongenetic sources of individual differences across a host of important developmental outcomes (e.g., aggression, depression, and social competence). These researchers seem to start conceptually with sibling differences in outcomes and then work backward to understand the experiences that produced such differences. Figure 2 shows a hypothetical distribution of scores on an outcome variable. It depicts possible links between several types of sibling differential experiences and sibling differences in an outcome variable. The key question is: Why are two children in the same family so different from one another (Dunn & Plomin, 1990)? As mentioned earlier, this approach is rooted in modern behavioral genetic theory (Plomin et al., 1997). Work in this area began with Plomin and Daniels' (1987) review article concerning the extent of sibling differences found in behavioral genetic studies. They pointed out that after one controls for genetic relatedness of the pairs, two children are as different as children picked at random with respect to personality and adjustment. Children in the same family, however, can differ for both genetic and nongenetic reasons. Studies using monozygotic twins showed that some sibling differences could not be explained by inherited genetic differences (Pike & Plomin, 1996). Plomin and Daniels (1987) hypothesized that the most important environmental factors in children's development must be experiences that children growing up in the same family do not share (see also Dunn & Plomin, 1990). A criticism of this approach has been that nonshared environment is just an anonymous component of variance and that behavioral geneticists need to measure the environment directly (e.g., Wachs, 1992). Almost a decade earlier, Rowe and Plomin (1981) proposed several areas of within-family differences that should be investigated (e.g., birth order and gender effects, differential treatment by parents, and asymmetrical sibling influences, and unique peer group experiences). Since then, most of the focus has been on directly uncovering environmental sources of sibling differences and understanding their role in individual differences in development.

An example of this work is the Nonshared Environment in Adolescent Development project (NEAD; Reiss et al., 2000). Using data from NEAD, Pike and colleagues (Pike, McGuire, Hetherington, Reiss, & Plomin, 1996) examined the relationship between parent-child negativity and children's externalizing and internalizing behavior problems. The study included pairs of monozygotic twins, dizygotic twins, and full siblings in never-divorced families and full siblings, half siblings, and biologically unrelated siblings in stepfamilies. This design allowed investigators to examine correlations between parent-child relationship dimensions (e.g., parent-child conflict) and children's adjustment (e.g., depression), just like most studies in the family relationship area. That is, the older siblings' (and the first twins') parent-child relationship scores were correlated with their outcomes and younger siblings' (and

FIGURE 2. Outcome-oriented nonshared environment researchers examine sources of sibling differences in outcomes in order to understand individual differences in development. C = child's score on an outcome measure. S = sibling's score on an outcome measure.

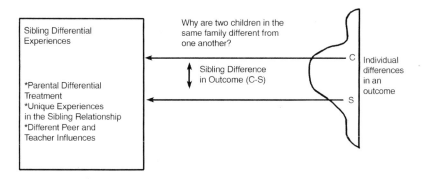

the second twins') parent-child relationship scores were correlated with their outcomes. The design, however, also permitted the researchers to take these analyses two steps further. They examined associations across the sibling dyad and across sibling type using sibling cross-correlations. A sibling cross-correlation is when the older siblings' scores for measure A are correlated with the younger siblings' scores for measure B (and the reverse). In a multivariate behavioral genetic model, the sibling cross-correlations are the key to disentangling the genetic and environmental contributions to the association between the parent-child measure and the adjustment measure. If the cross-correlations show the following pattern: MZ twins > DZ = Full sibling > half siblings > unrelated siblings, then the association between the two measures may be genetically mediated. For example, the correlation between parent-child conflict and children's aggressive behavior could be due to child effects with parents reacting to children's difficult temperament styles. If the sibling cross-correlations are high, but do not differ across sibling dyad type, then shared environmental contributions are implicated. Pike and colleagues (1996), however, found nonshared environmental mediation for some of the associations examined. This means that there was significant covariation between the parent-child measures and the adjustment measures that could not be accounted for by genetic or shared environmental contributions. The nonshared environment factors contributing to the association are not known and they could be due to methodological factors (e.g., rater bias specific to each child) or to real phenomena (e.g., family experiences specific to each child). This approach is not concerned with discovering sources of nonshared environment per se, but instead with understanding the etiology of the connections between measures of the parent-child relationship and children's development outcomes.

Other Distinctions

Recently, two other distinctions have been suggested for the understanding of nonshared environment: object versus effective nonshared environment, and single-system versus multiple-system processes (see Table 1). In their review, Turkheimer and Waldron (2000) differentiated between objective versus effective nonshared environment. *Objective Nonshared Environments* are observable differences in sibling environments. They point out that differences in objective environments may or may not lead to sibling differences in outcomes such as temperament or behavior problems. This is consistent with the experience-oriented approach to nonshared environment mentioned above. *Effective Nonshared Environments*, on the other hand, are experiences that produce sibling differences. This is consistent with the outcome-oriented approach discussed above. Turkheimer and Waldron (2000) added an additional level by arguing that situations producing sibling differences may or may not be objectively shared. They used the example of parental divorce. Each sibling may react differently to this shared family event, which would result in sibling differences in adjustment. They were criticizing Plomin and Daniels' (1987) statement that sibling differences in behavior could not stem from shared family characteristics or events. It is important to note that Plomin and Daniels' (1987) hypothesis is only one of several ideas being tested in the nonshared environment area. Still, Turkheimer and Waldron's (2000) description of these two different types of nonshared environment is very useful.

Reiss and colleagues (2000) also expanded the concept of nonshared environment by outlining single-system and multiple-system processes (see Table 1). In *single-system* contexts, sibling differences in experiences are linked to individual differences in adjustment, but the siblings are not engaging in the social comparison process. It has been argued that children learn via watching and comparing themselves to others (e.g., Bandura, 1977), including their siblings (e.g., Dunn, 1988). In this situation, however, the siblings are simply experiencing different social influences. They are not aware of the sibling's unique experiences or do not care enough to pay attention to them. For example, a parent may treat the siblings exactly the same when all three are together, but one child may get more attention or more time alone with the parent in a dyadic context. Hence, one child will have lower behavior problems because of the additional time spent with mom or dad and not because the child feels "favored" (obviously, genetic factors and direction-of-effects issues are being ignored in this example). The two siblings may not even realize that their parent is treating them differently.

Turkeimer and Waldron (2000) argued that single-system situations are already captured in traditional between-family socialization studies. For many decades, studies of links between parent-child relationships and children's adjustment focused on one child per family and compared children across (i.e., between) different families (see Collins, Maccoby, Steinberg, Hetherington, & Bornstein, 2000, for a review). The dyad was the unit of analysis. Between-family studies have shown that children differ in adjustment because of different interactions with their parents. Within-family studies (i.e., sibling studies) have found that siblings in the same family differ in adjustment because of different interactions with the parent. Turkheimer and Waldron (2000) asked: What is the difference? The answer is: There is little or no difference at the empirical level, but an important difference at the conceptual level. The findings will be interpreted very differently in the two research contexts. It is true that both between- and within-family studies could detect the association between a dimension of the parent-child relationship and an aspect of children's adjustment when single-system processes are in play. Between-family studies, however, could lead to the erroneous conclusion that the association was due exclusively to differences in parenting across families and not to differences in child characteristics or the nature of the dyad. This claim was made in early socialization studies, although it is rarely made today (Collins et al., 2000). Within-family studies showed that the parents often have unique relationships with each of their children. Sibling studies, even those focusing on single-system processes, contributed to our new appreciation for the role of child effects in development.

It is also important to reiterate that between-family designs do not permit researchers to disentangle shared from nonshared environmental influences on a behavior. For example, Figure 3 shows the results of a hypothetical between-family study and a within-family study. The first picture depicts an association between parents' scores on a parenting measure and children's scores on an outcome measure (e.g., depression). The overlap represents the extent to which differences in parenting covary with differences in children's adjustment across families. The second picture shows the overlap between parenting scores and older siblings' depression scores (i.e., NS_1), parenting scores and younger siblings' depression scores (i.e., NS_2), and overlap across all three measures (i.e., S). NS_1 and NS_2 represent links unique to each parent-child relationship within the family. S is the extent to which parenting is linked to sibling similarity in depression. Thus, within-family study designs allow researchers to examine both between-family and within-family differences.

Reiss and colleagues' (2000) second type of nonshared environment processes, *multiple-level* processes, can only be fully uncovered using within-family studies. In this situation, the siblings are knowledgeable about each other's experiences and are engaging in the social comparison process. Surprisingly, nonshared environment studies rarely measure the degree of social comparison within the dyad. Daniels and colleagues (Baker & Daniels, 1990; Daniels, 1986; Daniels & Plomin, 1985) did examine siblings' perceptions of the degree of sibling differential experiences. Teens were instructed to rate their relative experiences using a 5-point scale, using the Sibling Inventory of Differential Experiences (SIDE). The rating is analogous to the relative difference score mentioned above. If the child reported that the mother is "much more" strict with the sibling then the score was a 1 and a "little more" strict was a score of 2. They could report that she treats the siblings equally (a score of 3) or that she is more strict with the child being interviewed (a little more = a score of 4 and much more = a score of 5). This strategy for measuring differential experiences has been used in other studies (e.g., Dunn et al., 1990; McGuire, Dunn, & Plomin, 1995; McHale & Pawletko, 1992; Reiss et al., 2000). In contrast to difference scores created from two separate measures of the environment, this is a more direct measure of children's perceptions of sibling differential experiences. There are, however, several problems with studies that have used this index of within-family differences. First, the measure may not truly capture the social comparison process. It is also not known if children use information about sibling differential experiences when making sense of themselves or the world around them. The children may believe that such differences are fair, and so, they have little impact on their feelings and behaviors. Second, some researchers interviewed only one child per family. Relying on only one reporter does not provide a complete picture of the family situation. Third, many studies

FIGURE 3. Between-family designs only examine covariance between a parenting measure and one child's outcome measure (e.g., depression). Within-family designs examine the covariance between a parenting measure and both siblings' outcome measures. NS_1 = overlap unique to the parenting measure and the older siblings' depression scores. NS_2 = overlap unique to the parent measure and the younger siblings' depression score. S = overlap across all three scores.

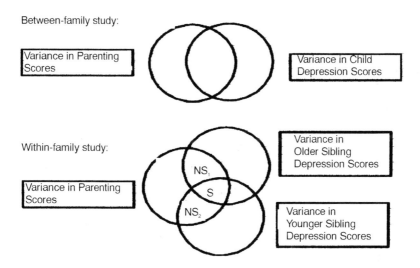

combine children's reports of nonshared environment with parent reports and/or observational measures, rather than focusing on children's awareness of the degree of differential experiences (e.g., Reiss et al., 2000). Even a recent article on the importance of sibling deidentification effects (Feinberg & Hetherington, 2000) relied exclusively on indirect evidence (i.e., sibling correlations for personality and adjustment across different age spacing groups). Still, several people have argued for the "reintroduction" of the concepts of sibling rivalry and sibling deidentification in social development (Feinberg & Hetherington, 2000; Reiss et al., 2000; Sulloway, 1996). Clearly, more work needs to be conducted concerning multi-system nonshared environment processes and their ties to children's socio-emotional adjustment and development.

WHERE IS THE NONSHARED
ENVIRONMENT FIELD GOING?

Books and reviews published in the nonshared environment area have already focused on the general issues in the field or have summarized and critiqued previous studies (Dunn & Plomin, 1990; Hetherington, Reiss, & Plomin, 1994; Hoffman, 1991; Pike & Plomin, 1996; Plomin & Daniels, 1987; Reiss et al., 2000; Rowe & Plomin, 1981; Turkheimer & Waldron, 2000). Consequently, the remainder of this paper will focus on areas that should be targeted for future research. Four topics gaining attention in the current nonshared environment literature are: (1) age-related changes and development; (2) the role of the self; (3) the role of context; and (4) the importance of extrafamilial experiences. Before turning to these topics, it is important to quickly review the general findings in the field. Over the last two decades, most work in the nonshared environment area has concerned parental differential treatment as opposed to differential experiences within the sibling relationship or outside the family. Studies have found that receiving more parental discipline and less warmth relative to one's sibling is related to higher levels of behavior problems and lower levels of self-esteem in children (e.g., Anderson, Hetherington, Reiss, & Howe, 1994; Conger & Conger, 1994; Dunn et al., 1990; McHale & Pawletko, 1992; McHale et al., 1995; McGuire et al., 1995; Mekos, Hetherington, & Reiss, 1996; Volling & Elins, 1998). Behavioral genetic studies show that associations between parent-child relationships and children's adjustment are also genetically mediated (e.g., Reiss et al., 2000). In fact, once genetic factors are taken into account, "pure" nonshared environmental contributions to links between parenting and children's adjustment appear to be minimal. This has led some scientists to question whether or not parental differential treatment is really a source of nongenetic contributions to individual differences in development (Reiss et al., 2000; Turkheimer & Waldron, 2000). This conclusion may be more troubling to outcome-oriented researchers than it is to experience-oriented researchers, who view nonshared processes as a product of interactions between the person and the environment. Therefore, it may not be surprising that experience-oriented researchers have begun focusing on the issue of the context of parental differential treatment and outcome-oriented researchers have started examining sources of nonshared environment outside the family. Scientists from both approaches, however, are concerned with age-related and developmental changes in nonshared environment and their relevance for children's behavior and adjustment.

Age-Related Changes and Development

Understanding age-related changes in siblings' unique experiences is difficult for at least three reasons. First, the field is relatively new, and so data are incomplete at each stage of the lifespan. Not surprisingly, most studies have focused on childhood and adolescence (see Turkheimer & Waldron, 2000). More research is needed on siblings at every development stage, but especially during infancy (see Braungart, Fulker, & Plomin, 1992; McGuire & Roch-Levecq, in press; Saudino & Plomin, 1997; Volling & Elins, 1998 for exceptions) and adulthood (see Baker & Daniels, 1990; Bouchard & McGue, 1990; McGonigle, Smith, Benjamin, & Turner, 1993; Vernon, Jang, Harris, & McCarthy, 1997 for exceptions). Second, the only dimension of nonshared environment that appears across every age group is maternal differential discipline/control. Studies vary greatly in terms of the other nonshared environment dimensions included in the analysis. Some investigations have included other parental differential treatment measures such as differential favoritism (e.g., McHale, Updegraff, Jackson-Newman, Tucker, & Crouter, 2000; Volling & Ellins, 1997), differential responsiveness and warmth (e.g., Brody, Stoneman, & McCoy, 1992; McGuire et al., 1995) and differential expectations (e.g., Daniels, Dunn, Furstenberg, & Plomin, 1985). Other investigators have broadened their measures of within-family differences to include family structure characteristics (e.g., birth order; Rodgers & Rowe, 1985), unique experiences within the sibling relationship (e.g., Dunn et al., 1990), differential experiences with friends and peers (e.g., Baker & Daniels, 1990; Daniels, 1986; Pike & Plomin, 1997), and specific life events (e.g., Eley & Stevenson, 2000). Still, it is nearly impossible to compare studies across ages when researchers are focusing on separate dimensions and are employing several measures. In addition, few reports include descriptive data about the extent of sibling differential experiences in the family, making direct comparisons even more difficult. Third, it is possible that there will not be clear, uniform changes in the unique experiences siblings encounter across development. Siblings' distinct experiences depend greatly on the personal characteristics of the children involved and the dynamics of the family system. Exploring mean level changes and percentages across development may not be useful. For example, Turkheimer and Waldron (2000) examined the effect sizes for nonshared environment contributions to adjustment across three age groups and found larger effects in childhood and adulthood compared to adolescence. These results are difficult to interpret because, in addition to the factors mentioned above, nonshared environment processes may change across development. The authors do discuss these issues, but it is also important that nonshared environment researchers consider them as well. It may be more advantageous for us to explore how developmental processes interact with siblings' nonshared experiences than to examine age-related changes at the global level.

Several investigators have said that one of the reasons parents treat (non-twin) siblings in the same family differently is because the children are at different developmental stages (e.g., Dunn et al., 1990; McHale et al., 2000; Volling, 1997). Volling (Volling, 1997; Volling & Elins, 1998) calls this normative parental differential treatment and differentiates it from parental favoritism. (This is similar to the distinction made by Reiss et al., 2000, between single-system and multi-system nonshared environment processes). It is hard to imagine that a parent would expect the same degree of obedience from a one-year-old and her three-year-old sibling brother. A mother may have to discipline her toddler more than her infant sibling, but that does not mean the mother favors the infant over the older child. Volling also argues that the extent to which parental differential treatment is based on developmental differences may change across the lifespan. Parental expectations, for instance, may differ greatly when the siblings are one and three years old, but less so when they are 11 and 13 years old, and maybe not at all when they are 21 and 23 years old. In addition, normative parental differential treatment may not be linked to children's temperament or adjustment because it is an appropriate parenting strategy; a strategy that takes into consideration each child's maturity level (Volling, 1997). One important avenue for future research will be to understand when parental differential treatment is based on age-related factors, which should change across time, and when it is based on parents' reactions to children's personality traits and/or difficult behavior, which may not change across time.

The Role of the Self

Another important topic to explore is the role of the developing self in nonshared processes. Two components of the self-system may be particularly relevant to the study of nonshared environment: children's self-perceptions of differential treatment and children's identity development. Reiss et al. (2000) suggest that the psychoanalytic perspective could provide a framework for understanding nonshared environment because the theory focuses on people's unique representations of themselves and others. The key is not the events per se, but how the developing siblings *perceive* the situation. This is consistent with W. I. Thomas' (1928) assertion that subjective experiences have real consequences. A study by McHale and colleagues (2000), rooted in this constructivist perspective, found that siblings' perceptions of fairness were associated with their self-esteem to a greater degree than more objective measures of parental differential treatment. Kowal and Kramer (1997) also highlighted the issue of fairness in their study of school-aged children; they found that 75% of the children who recognized the presence of parental differential treatment in their families also reported they did not think the situation was unfair. The children took into consideration personal differences (e.g., age and personality factors) and circumstances when discussing the appropriateness of their parents' behavior. The sibling pair's shared perception of the situation may also be relevant for understanding nonshared environmental processes. Sibling agreement about parental differential treatment appears to increase from middle childhood to adolescence and adulthood (Baker & Daniels, 1990; Daniels & Plomin, 1985; Pike, Manke, Reiss, & Plomin, 2000). This has led some to speculate that dyads develop a more shared schema about sibling differences and family roles as they grow older (Pike et al., 2000). A topic for future research is the role of children's perceptions of fairness and their interpretations of the differences between them and their siblings. It is tempting to conclude that such processes will not come into play until middle childhood when most children have a fairly sophisticated sense of self (Harter, 1983). Dunn's work on social-cognitive development during early childhood, however, has shown us that even very young children monitor their older siblings' behavior (e.g., Dunn, 1988). Children may be aware of sibling differences from a young age, even if they have trouble articulating it to others.

Identity development, on the other hand, is traditionally associated with development during middle childhood and adolescence. Interestingly, siblings were highlighted as significant players in identity formation some twenty years ago (e.g., Schachter et al., 1978; Tesser, 1980). In a recent paper, Feinberg and Hetherington (2000) argued that sibling identification theory has a place in current sibling studies. They review the basic principles underlying this approach to understanding identity development. Specifically, sibling deidentification theory states that children actively differentiate themselves from their siblings, especially in pairs that contain children with very similar personal characteristics (e.g., close in age or the same gender). Sibling rivalry is considered normative and the deidentification process is a way for children to deal with this painful situation. It allows children to avoid directly competing for parental love and attention, to defend against unwanted comparisons with siblings, and to create unique identities within the family. Aside from Schachter's own work (e.g., Schachter, Shore, Feldman-Rotman, Marquis, & Campbell, 1976; Schachter et al., 1978), the theory gained little empirical attention during the 1980s and 1990s. This may be due to measurement difficulties and to the focus on sibling designs in behavioral genetics. Feinberg and Hetherington (2000) argue that such processes should to be taken into consideration and direct tests of the theory need to be incorporated into future work on sibling differences.

The Role of Context

The role of context in development is receiving increasing attention in several fields, including personality development and developmental psychopathology (Cicchetti & Aber, 1998; Rubin, 1998). It is an interesting direction for future work in the nonshared environment area as well. There is evidence that "process-context interactions" (Bronfenbrenner & Crouter, 1983) exist in relation to parental differential treatment. For example, Quittner and Opipari (1994) found that the magnitude of maternal differential treatment was greater in dyads where the younger sibling has cystic fibrosis compared to pairs that contain two healthy children. This is not surprising. The difference in parental treatment, however, was still greater in families with an ill sibling even after controlling for the amount of time the mother spent with the younger siblings in medical care situations. This finding is consistent with the study by McHale and Pawletko (1992) mentioned above; it showed greater differential treatment in families with a disabled child compared to control families. In addition, the associations between parental differential treatment and children's adjustment differed in the two family contexts. Older siblings who spent more time with their mothers compared to their disabled siblings (in certain activities) showed higher levels of depression and anxiety compared to older siblings in the control sample. This is the opposite pattern typically found in the nonshared environment literature; usually more time spent with a parent relative to the sibling is associated with more positive outcomes.

Data from normative samples also support the idea that siblings' differential experiences are not uniform across family conditions. Using data from the NEAD project and a family systems perspective, O'Connor and colleagues (1998) found meaningful parental differential effects in some, but not all families. Mekos et al. (1996) found differences between remarried and nondivorced families in the same sample. McGuire, Clifford, and Fink (2001) investigated links between parental differential treatment and children's behavior problems in two potentially problematic family circumstances (i.e., families high in conflict or low in warmth) compared to other situations. Many of their results supported an additive risk model. Specifically, children who were treated less favorably compared to siblings and lived in difficult family contexts had the highest levels of behavior problems. Children who receive equal treatment in families high in warmth or low in conflict had the lowest scores, and all other combinations were in the middle range.

It is important that nonshared environment researchers explore the conditions surrounding sibling differential experiences. The majority of the work on correlates of parental differential treatment has focused on the role of children's personal characteristics (e.g., temperament; Brody et al., 1987). Parental differential treatment does appear to be associated with other negative family experiences, such as parental stress (Crouter, McHale, & Tucker, 1999) and marital discord (Deal, 1996). This may be true, however, only during specific developmental time periods and only in community-based samples. Studies are needed that attempt to understand how links between nonshared environment and children's adjustment changes across diverse family conditions and even across cultures and historical time periods. In some countries, sibling differential experiences are dictated by society (i.e., the firstborn receives all the land). It will be interesting to see if results found using samples from the United States generalize to other cultures. Little research on nonshared environment has been conducted outside the U.S. (see Boer, 1992 for an exception) or Western society.

The Importance of Extrafamilial Experiences

It is crucial that researchers also move the concept of nonshared environment outside the family. Since the beginning, investigators have argued that siblings' unique experiences with peers and teachers will be a significant source of nonshared environment (e.g., Rowe & Plomin, 1981; Dunn & Plomin, 1990). Turkheimer and Waldron (2000) found that studies including sibling differences in extrafamilial experiences had larger effect sizes than those focusing exclusively on family dynamics. Yet, only a handful of sibling studies have incorporated measures of each sibling's friendships, peer relations, and teacher-child interactions. Everyone would agree that the playground, neighborhood, and school are rich contexts to search for influences tied to sibling individuation. In fact, Harris (1998) argues that peers are the primary source of environmental influences on children's personality development. Why are there so few studies examining extrafamilial environments? It makes sense conceptually that nonshared environment researchers would look inside the family first for sources of unique experiences, especially if the model includes processes such as sibling rivalry and sibling social comparison. The home is a place where siblings watch each other grow up and, sometimes, compare each other's experiences. There is also a methodological reason why nonshared environment has stayed in the family. Measuring sibling differential experiences outside of the home will be very difficult. Studies to date have relied on parent reports and children's self-reports of children's extrafamilial experiences. Still, the results of this research show that nonshared environment investigators need to explore these sources of individual differences.

Studies of school-aged and adolescent siblings have found that children report greater sibling differences in peer group experiences than in parental treatment (e.g., Daniels & Plomin, 1985; Pike et al., 2000). Nonshared environment estimates are higher for measures of friendship than for parent-child and sibling relationships (Pike & Plomin, 1997). It cannot be assumed, however, that extrafamilial experiences are pure environmental measures of within-family differences. Behavioral genetic studies reveal that parent reports and children's self-reports of peer group characteristics and peer popularity are heritable (Manke, McGuire, Reiss, Hetherington, & Plomin, 1995; McGuire et al., 1999). Sibling reports of their differential peer experiences are linked to sibling differences in temperament characteristics, especially sociability (Baker & Daniels, 1990; Daniels, 1986). Pike and Plomin (1997), however, interviewed MZ (identical) twins about the sources of their behavioral differences (MZ twins cannot differ because of inherited genetic differences). Many twins cited their unique experiences with friends as a significant source of twin differences. Other types of unique events included unique romantic experiences, different teachers at school, differential reactions to shared life events (e.g., death of a close relative), and involvement in separate after-school activities. It is clear that much of the action in the nonshared environment arena will be outside of the family context. The trick will be to find ways to capture these events. It may be time for nonshared environment researchers to reach out to scientists studying peer and teacher-child relationships and propose joint studies or to suggest adding a sibling model to already existing investigations of extrafamilial relationships.

Other Issues. Thus far, this review has not covered two other sources of nonshared environment in depth: unique experiences within the sibling relationship itself and specific (possibly unsystematic) processes. Siblings often provide distinct perspectives concerning their "shared" relationship (Dunn & Plomin, 1990; McGuire, Manke, Eftekhari, & Dunn, 2000). This will be an important avenue for future research on within-family differences. The most interesting work on the role of sibling relationships, however, may concern their joint experiences. The NEAD study showed significant shared environmental mediation between measures of the sibling relationship and children's adjustment during adolescence (Reiss et al., 2000). Over 15 years ago, Patterson (1984) argued that some conduct-disordered boys "train" their brothers to be antisocial through hostile sibling interaction. Rowe and Gulley (1992) found that brother-brother pairs in warm sibling relationships were more similar in delinquency. They assert that siblings may "recruit" each other to participate in criminal acts. A recent study also suggests a link between sibling relationship characteristics and sibling similarity in delinquency (Slomkowski, Rende, Conger, Simons, & Conger, 2001). Future research is needed to understand under what conditions the sibling relationship promotes the development of psychopathology.

Other potentially important sources of sibling differences in development are life events and unsystematic experiences. Some have argued that this type of nonshared environment may be so random that it may be nearly impossible to investigate (e.g., Plomin & Daniels, 1987; see also Turkheimer & Waldron, 2000). A recent twin study of threat and loss shows us that some of these experiences can be examined using a sibling design (Eley and Stevenson, 2000). Sixty-one pairs of twins were included in which at least one child had significantly high depression or anxiety scores. The investigators found that anxious twins experienced a higher number of threatening events (e.g., personal challenges, trauma, and physical jeopardy) over the last year compared to their normal co-twins. This difference was still significant when MZ twin pairs were the only twins included in the analysis. In addition, depressed twins were higher in friendship and family relationships problems compared to their normal co-twins; the difference was significant when only MZ twin pairs were examined. Still, many have pointed out that hard-to-study unsystematic or nonlinear events may accumulate over time leading to meaningful sibling differences in development (e.g., Dunn & Plomin, 1990; Plomin & Daniels, 1987; Reiss et al., 2000; Turkeimer, 2000; Turkheimer & Waldron, 2000). Furthermore, some sibling differences may be due to higher-order genetic-environment interactions that should not be categorized as nonshared "environmental" events at all (see Molenaar, Boomsma, & Dolan, 1993; Turkheimer, 2000; Turkheimer & Gottesman, 1996). Discovering and documenting these complicated processes will be challenging, but it will be important to understand their part in children's adjustment.

CONCLUSION

Nonshared environment research is being approached from at least two different perspectives. One group of researchers focuses on differential experiences within the family, while another concentrates on uncovering environmental sources of sibling and individual differences in outcomes. Work from both "camps" is progressing toward understanding the complicated nature of nonshared environment processes and how they change across time and context. For instance, sometimes shared events lead to sibling differences and other times nonshared events lead to sibling similarities. In addition, nonshared environment may involve siblings experiencing different environments or entail siblings engaging in the social comparison process. Future work should explore developmental changes, contextual factors, and extrafamilial sources of nonshared environment. Work in these areas has been plagued by methodological and conceptual problems; still, several themes have emerged. This issue of parental differential treatment as a normative parenting style versus as parental favoritism requires more attention. The children's own perspectives and their notions of fairness should be examined. The uncharted territory of sibling differential experiences in school, on the playground, and around the neighborhood needs to be explored. Conceptual models that include these factors will give us a more complete picture of the role of nonshared experiences in sibling differences and in children's development.

REFERENCES

Anderson, E. R., Hetherington, E. M., Reiss, D., & Howe, G. (1994) Parents' nonshared treatment of siblings and the development of social competence during adolescence. *Journal of Family Psychology, 8*, 303-320.

Baker, L. A., & Daniels, D. (1990). Nonshared environmental influences and personality differences in adult twins. *Journal of Personality and Social Psychology, 58*, 103-110.

Bandura, A. (1977). *Social learning theory.* Englewood Cliffs, NJ: Prentice Hall.

Bernard, J. (1972). *The future of marriage.* NY: Bantam Books.

Boer, F. (1990). *Sibling relationships in middle childhood: An empirical study.* Leiden: DSWO Press.

Bouchard, T. J., & McGue, M. (1990). Genetic and rearing environmental influences on adult personality: An analysis of adopted twins reared apart. *Journal of Personality, 58,* 263-292.

Bowen, M. (1978). *Family therapy in clinical practice.* New York: Jason Aronson.

Braungart, J. M., Fulker, D. W., & Plomin, R. (1992). Genetic mediation of the home environment during infancy: A sibling adoption study of the home. *Developmental Psychology, 28,* 1048-1055.

Brody, G. H., Stoneman, Z., & Burke, M. (1987). Child temperaments, maternal differential behavior, and sibling relationships. *Developmental Psychology, 23,* 354-362.

Brody, G. H., Stoneman, Z., & McCoy, J. K. (1992). Associations of maternal and paternal direct and differential behavior with sibling relationships: Contemporaneous and longitudinal analyses. *Child Development, 63,* 82-92.

Bronfenbrenner, U., & Crouter, A. C. (1983). The evolution of environmental models in developmental research. In P. Mussen (Eds.), *Handbook of Child Psychology* (pp. 358-414). NY: Wiley.

Cicchetti, D., & Aber, L. (1998). Contextualism and developmental psychopathology. *Development and Psychopathology, 10,* 137-141.

Collins, W. A., Maccoby, E. E., Steinberg, L., Hetherington, E. M., & Bornstein, M. H. (2000). Contemporary research on parenting: The case for nature and nurture. *American Psychologist, 55,* 218-232.

Conger, K. J., & Conger, R. D. (1994). Differential parenting and change in sibling differences in delinquency. *Journal of Family Psychology, 8,* 287-302.

Crouter, A. C., Manke, B. A., & McHale, S. M. (1995). The family context of gender intensification in early adolescence. *Child Development, 66,* 317-329.

Crouter, A. C., McHale, S. M., & Tucker, C. J. (1999). Does stress exacerbate parental differential treatment of siblings? A pattern-analytic approach. *Journal of Family Psychology, 13,* 286-299.

Daniels, D. (1986). Differential experiences of siblings in the same family as predictors of adolescent personality differences. *Journal of Personality and Social Psychology, 51,* 339-346.

Daniels, D., Dunn, J., Furstenberg, G., & Plomin, R. (1985). Environmental differences within the family and adjustment differences within pairs of adolescent siblings. *Child Development, 56,* 764-774.

Daniels, D. & Plomin, R. (1985). Differential experiences of siblings in the same family. *Developmental Psychology, 21,* 747-760.

Deal, J. E. (1996). Marital conflict and differential treatment of siblings. *Family Process, 35,* 333-346.

Dunn, J. (1988). *The beginnings of social understanding.* Cambridge, MA: Harvard University Press.

Dunn, J., & Plomin, R. (1990). *Separate lives: Why siblings are so different.* New York: Basic Books.

Dunn, J., Stocker, C., & Plomin, R. (1990). Nonshared experiences within the family: Correlates of behavior problems in middle childhood. *Development and Psychopathology, 2,* 113-126.

Eley, T. C., & Stevenson, J. (2000). Specific life events and chronic experiences differentially associated with depression and anxiety in young twins. *Journal of Abnormal Child Psychology, 28,* 383-394.

Feinberg, M. E., & Hetherington, E. M. (2000). Sibling differentiation in adolescence: Implications for behavioral genetic theory. *Child Development, 71,* 1512-1524.

Gottlieb, G., Wahlsten, D., & Lickliter, R. (1998). The significance of biology for human development: A developmental psychobiological systems view. In R. M. Lerner (volume Ed.), *Handbook of child psychology: Vol. 3, theoretical models of human development* (pp. 233-273). New York: Wiley.

Harris, J. R. (1998). The nurture assumption: Why children turn out the way they do. New York: Free Press.

Harter, S. (1983). Developmental perspectives on the self-system. In E. M. Hetherington, (Ed.), P. H. Mussen (Series Ed.), *Handbook of child psychology: Vol. 4, Socialization, personality, and social development* (pp. 275-385). NY: Wiley.

Hetherington, E. M., Reiss, D., & Plomin, R. (1994). *Separate worlds of siblings: The impact of nonshared environment on development.* Hillsdale, NJ: Erlbaum.

Hoffman, L. W. (1991). The influence of the family environment on personality: Accounting for sibling differences. *Psychological Bulletin, 110,* 187-203.

Kowal, A., & Kramer, L. (1997). Children's understanding of parental differential treatment. *Child Development, 68,* 113-126.

Manke, B., McGuire, S., Reiss, D., Hetherington, E.M., & Plomin, R. (1995). Genetic contributions to adolescents' extrafamilial interactions: Teachers, best friends, & peers. *Social Development, 44,* 238-256.

McGonigle, M. M., Smith, T. W., Benjamin, L. S., & Turner, C. W. (1993). Hostility and nonshared family contexts: A study of monozygotic twins. *Journal of Research in Personality, 27,* 23-34.

McGuire, S., Clifford, J., & Fink, J. (2001). Parental differential treatment in different family contexts: Testing multiple risk models. Manuscript submitted for publication.

McGuire, S., Dunn, J., & Plomin, R. (1995). Maternal differential treatment of siblings and children's behavioral problems: A longitudinal study. *Development & Psychopathology, 7,* 515-528.

McGuire, S., Manke, B., Eftekhari, A., & Dunn, J. (2000). Children's perceptions of sibling conflict during middle childhood: Issues and sibling (dis)similarity. *Social Development, 9,* 173-190.

McGuire, S., Manke, B., Saudino, K. J., Reiss, D., Hetherington, E. M., & Plomin, R. (1999). Perceived competence and self-worth during adolescence: A longitudinal behavioral genetic study. *Child Development, 70,* 1283-1296.

McGuire, S., & Roch-Levecq, A. C. (In press). Mothers' perceptions of differential treatment of infant twins. In R. N. Emde & J. K. Hewitt (Eds.) *The transition from infancy to early childhood: Genetic and environmental influences in the MacArthur Longitudinal Twin Study.* New York: Oxford University Press.

McHale, S. M., Crouter, A. C., McGuire, S., & Updegraff, K. A. (1995). Congruence between mothers' and fathers' differential treatment of siblings: Links with family relations and children's well being. *Child Development, 66*, 116-128.

McHale, S. M., & Pawletko, T. M. (1992). Differential treatment in two family contexts. *Child Development, 63*, 68-81.

McHale, S. M., Updegraff, K. A., Jackson-Newsom, J., Tucker, C. J., & Crouter, A. C. (2000). When does parents' differential treatment have negative implications for siblings? *Social Development, 9*, 149-172.

Mekos, D., Hetherington, E. M., & Reiss, D. (1996). Sibling differences in problem behavior and parental treatment in nondivorced and remarried families. *Child Development, 67*, 2148-2165.

Molenaar, P. C. M., Boomsma, D. I., & Dolan, C. V. (1993). A third source of developmental differences. *Behavioral Genetics, 23*, 519-524.

O'Connor, T. G., Hetherington, E. M., & Reiss, D. (1998). Family systems and adolescent development: Shared and nonshared risk factors in nondivorced and remarried families. *Development and Psychopathology, 10*, 353-375.

Patterson, G. (1984). Siblings: Fellow travelers in coercive family process. In R. Blanchard & D. Blanchard (Eds.) *Advances in the study of aggression: Vol. 1* (pp. 173-215). Orlando, FL: Academic Press.

Pike, A., Manke, B., Reiss, D., & Plomin, R. (2000). A genetic analysis of differential experiences of adolescent siblings across three years. *Social Development, 9*, 96-114.

Pike, A., McGuire, S., Hetherington, E.M., Reiss, D., & Plomin, R. (1996). Family environment and adolescent depressive symptoms and antisocial behavior: A multivariate genetic analysis. *Developmental Psychology, 32*, 590-603.

Pike, A., & Plomin, R. (1996). Importance of nonshared environmental factors for childhood and adolescent psychopathology. *Journal of the American Academy of Child & Adolescent Psychiatry, 5*, 560-570.

Pike, A., & Plomin, R. (1997). A behavioral genetic perspective on close relationships. *International Journal of Behavioral Development, 21*, 647-667.

Pike, A., Reiss, D., Hetherington, E. M., & Plomin, R. (1996). Using MZ differences in the search for nonshared environmental effects. *Journal of Child Psychology & Psychiatry & Allied Disciplines, 37*, 695-704.

Plomin, R., & Daniels, D. (1987). Why are children in the same family so different from each other? *The Behavioral and Brain Sciences, 10*, 1-16.

Plomin, R., DeFries, J. C., McClearn, G. E., & Rutter, M. (1997). *Behavioral genetics* (3rd ed.). New York: W. H. Freeman.

Quittner, A. L., & Opipari, L. C. (1994). Differential treatment of siblings: Interview and diary analyses comparing two family contexts. *Child Development, 65*, 800-814.

Reiss, D., & Oliveri, M. E. (1983). The family's construction of social reality and its ties to its kin network: An exploration of causal direction. *Journal of Marriage & the Family, 45*, 81-91.

Reiss, D., Neiderhiser, J., Hetherington, E. M., & Plomin, R. (2000). *The relationship code.* Cambridge, MA: Harvard University Press.

Rodgers, J. L., & Rowe, D. C. (1985). Does contiguity breed similarity? A within-family analysis of nonshared sources of IQ differences between siblings. *Developmental Psychology, 21*, 743-746.

Rowe, D. C., & Gulley, B. (1992). Sibling effects on substance abuse and delinquency. *Criminology, 30*, 217-233.

Rowe, D., & Plomin, R. (1981). The importance of nonshared (E1) environmental influences in behavioral development. *Developmental Psychology, 17*, 517-531.

Rubin, K. (1998). Social and emotional development from a cultural perspective. *Developmental Psychology, 34*, 611-615.

Saudino, K. J., & Plomin, R. (1997). Cognitive and temperamental mediation of genetic contributions to the home environment during infancy. *Merrill-Palmer Quarterly, 43*, 1-23.

Schachter, F.F., Gilutz, G., Shore, E., & Adler, M. (1978). Sibling de-identification judged by mothers: Cross-validation and developmental studies. *Child Development, 49*, 543-546.

Schachter, F.F., Shore, E., Feldman-Rotman, S., Marquis, R.E., & Campbell, S. (1976). Sibling deidentification. *Developmental Psychology, 12*, 418-427.

Siegel, M. (1987). Are sons and daughters treated more differently by fathers than by mothers? *Developmental Review, 7*, 183-209.

Slomkowski, C., Rende, R., Conger, K. J., Simons, R. L., & Conger, R. D. (2001). Sisters, brothers, and delinquency: Evaluating social influence during early and middle adolescence. *Child Development, 72*, 271-283.

Sulloway, F. J. (1996). *Born to rebel: Birth order, family dynamics, and creative lives.* New York: Pantheon Books.

Tesser, A. (1980). Self-esteem maintenance in family dynamics. *Journal of Personality and Social Psychology, 39*, 77-91.

Thomas, W. I., & Thomas, D. S. (1928). *The child in America.* NY: Alfred P. Knopf.

Turkheimer, E. (2000). Three laws of behavior genetics and what they mean. *Current Directions in Psychological Science, 9*, 160-164.

Turkheimer, E., & Gottesman, I. L. (1996). Simulating the dynamics of genes and environment in development. *Development & Psychopathology, 8*, 667-677.

Turkheimer, E., & Waldron, M. (2000). Nonshared environment: A theoretical, methodological, and quantitative review. *Psychological Bulletin, 126*, 78-108.

Vernon, P. A., Jang, K. L., Harris, J. A., & McCarthy, J. M. (1997). Environmental predictors of personality differences: A twin and sibling study. *Journal of Personality and Social Psychology, 72*, 177-183.

Volling, B. L. (1997). The family correlates of maternal and paternal perceptions of differential treatment in early childhood. *Family Relations, 46*, 227-236.

Volling, B. L., & Elins, J. L. (1998). Family relationships and children's emotional adjustment as correlates of maternal and paternal differential treatment: A replication with toddler and preschool siblings. *Child Development, 69*, 1640-1656.

Wachs, T. D. (1992). *The nature of nurture.* Newbury Park, CA: Sage.

White, L., & Brinkerhoff, D. (1981). The sexual division of labor: Evidence from childhood. *Social Forces, 60*, 170-181.

From Behavioral Genetics
to Molecular Genetics:
Direct Tests of Genetic Hypotheses
for Behavioral Phenotypes

Thalia C. Eley

SUMMARY. The aim of this paper is to show how molecular genetics can be brought into mainstream psychological studies, using genes as directly measured variables. The first section of the paper concentrates on the basic methods and issues involved in molecular genetic work including genetic markers, linkage and association. This is followed by an overview of the more exciting results from the field to date. These are organized around the five areas of hypotheses that are outlined in the accompanying paper on behavioral genetic methodology (Petrill, this volume). The first set of hypotheses relate to multivariate issues, the second to longitudinal analyses, third are group effects, most notably sex effects, fourth are hypotheses considering the genetic relations between abnormal and normal development, and the fifth is interaction with the environment. While it is unlikely that many social scientists will join the hunt for genes associated with behaviors, that does not preclude them

Thalia C. Eley is affiliated with the Social, Genetic and Developmental Psychiatry Research Centre, Institute of Psychiatry, King's College London, De'Crespigny Park, London SE5 8AF, UK (E-mail: t.eley@iop.kcl.ac.uk).

The author is funded by a Research Training Fellowship from the UK Medical Research Council.

[Haworth co-indexing entry note]: "From Behavioral Genetics to Molecular Genetics: Direct Tests of Genetic Hypotheses for Behavioral Phenotypes." Eley, Thalia C. Co-published simultaneously in *Marriage & Family Review* (The Haworth Press, Inc.) Vol. 33, No. 1, 2001, pp. 57-75; and: *Gene-Environment Processes in Social Behaviors and Relationships* (ed: Kirby Deater-Deckard, and Stephen A. Petrill) The Haworth Press, Inc., 2001, pp. 57-75. Single or multiple copies of this article are available for a fee from The Haworth Document Delivery Service [1-800-HAWORTH, 9:00 a.m. - 5:00 p.m. (EST). E-mail address: getinfo@haworthpressinc.com].

57

from using such genes in studies of social behavior and relationships (Plomin & Rutter, 1998). *[Article copies available for a fee from The Haworth Document Delivery Service: 1-800-HAWORTH. E-mail address: <getinfo@haworthpressinc.com> Website: <http://www.HaworthPress.com> © 2001 by The Haworth Press, Inc. All rights reserved.]*

KEYWORDS. Molecular genetics, methodology, development, gene-environment interaction

Traditional molecular genetic research has concentrated on single gene disorders, such as Huntington's disease, where one gene is both necessary and sufficient to cause the disease. As a result, the influence of genes is generally thought of in this way, and those who study variations in normal behaviour and development may not see the relevance of genetic research to the phenotypes they study. However, as is clear from the behavioural genetic research on complex traits, such as that covered in this volume, these phenotypes do not have a simple pattern of Mendelian inheritance. In contrast to single-gene disorders, complex behaviours and disorders are influenced by a wide array of factors, both genetic and environmental. As such, individual genes may be associated with individual differences in a wide variety of complex traits and behaviours. Molecular genetic techniques have therefore had to develop beyond strategies suitable for studying single gene disorders, towards methods that recognise that individual genes do not operate alone, but rather in conjunction with the environment and with other genes.

Several techniques within the field of behavioural genetics have been developed to test differing genetic hypotheses (see Petrill, this volume), and these provide valuable insights into the role of genetic and environmental influences on variance in measured phenotypes. Such data can inform further research, for example, by identifying whether a certain trait is genetically linked to or distinct from another. Molecular genetics offers a complementary approach to understanding the aetiology of complex traits, providing techniques to establish the role of specific individual genes. There is a long and productive history of psychological studies of specific features of the environment involved in the development of behaviour and social relationships. However, the role of specific genes in such development has been much less widely explored, and is now ripe for exploration. Fortunately, while behavioural genetic designs require specific sample types (e.g., twins or adopted families), individual genes may be included into regular studies and, without too much difficulty, analysed alongside other variables of interest.

The genes involved in complex traits are necessarily of smaller effect size (being neither necessary nor sufficient to cause disorder) than those for single-gene disorders and are known as *Quantitative Trait Loci* (QTLs). Quantitative Trait Loci are genes involved in individual differences for any quantitatively measured trait, at the extreme end of which there may be a group of behaviours classified as a disorder. As such QTLs are important not only for understanding the aetiology of complex traits, but also for understanding complex disorders such as those found in the field of psychology and psychiatry. The demands of identifying QTLs are very different from identifying a gene that is the single cause of a disorder, and different methods have therefore evolved. There are two main strategies used to identify gene-trait or gene-disorder relationships within molecular genetics: *linkage* and *association*. Both these approaches make use of genetic *markers*. The remainder of this paper is divided into two broad sections: methods and results. The methods section provides a brief overview of the use of genetic markers before moving on to outline association and linkage methods. The results section provides an overview of some of the most promising results to date.

MOLECULAR GENETIC METHODS

Genetic Markers

Molecular genetic approaches are now based on the use of the vast number of genetic markers that have been identified within the human genome. A marker is a stretch of DNA that varies between individuals in a population (McGuffin, Owen, O'Donovan, Thapar, & Gottesman, 1994). Different versions of a marker are called *alleles*. A technique called Polymerase Chain Reaction (PCR) is used to ascertain the *genotype* (pair of alleles–one from each of the two copies of the chromosome of interest) for any one individual. PCR has revolutionized molecular genetics as it makes it possible to produce large numbers of copies of the region of interest from a very small amount of template DNA. Having produced such a large quantity of the section of DNA of interest this can then be analyzed in order to establish the genotype for that marker for each individual. There are several different types of markers, but these can be grouped into two main types. The first group of markers are those in which a section of DNA has mutated in some way resulting in a small change between the original and new sections. The two main types of marker in this are Restriction Fragment Length Polymorphisms (RFLPs) and Single Nucleotide Polymorphisms (SNPs). As there are only two alleles for these markers (the original and the mutated version), one of which is usually rather rare, they are not all that informative. The second group of markers are multi-allelic, i.e., there are several differing versions of the marker found in the population. These most commonly consist of markers in which a section of DNA has repeated a varied number of times. The repeats can be anything from 2 to 100 base-pairs long. Repeats of only 2 to 4 base-pairs are known as short sequence repeats (SSRs), whereas longer repeats are known as Variable Number Tandem Repeat markers (VNTRs). These repeat markers are sometimes found within the protein-coding regions of a gene and thus may have a functional role. For example, one allele may result in increased transcription of an enzyme compared to the other allele(s).

These markers have one main feature in common: they vary within the population. These variations occur both within a family (different members of a family will have varying genotypes for each marker) and within a whole population. As such, variations within these markers can be analyzed both within families (using linkage) and within populations (association).

Linkage

Linkage is the co-segregation of a marker with a disease within families. In other words, if a certain version of a stretch of DNA is more commonly found to be present in those members of a family carrying the disease, than those who do not carry the disease, then the marker and disease are said to be linked (Plomin, DeFries, McClearn, & Rutter, 1997). The marker may not itself be involved in the etiology of the disorder, but may be nearby and therefore be inherited along with the disease causing allele. The further away the marker is from the disease gene, the less strong the relationship between the disease and the marker, resulting in lower power to detect the true linkage. This is how the genes for single-gene disorders such as Huntington's disease were identified. In such a disease, the disease-causing gene is always present in those with the disease, and always absent in those without the disease (assuming they have passed through the age of risk). As such, the gene has a very clear and strong link to the disease, which can be traced down through a large family pedigree.

In contrast, for traits where many genes are involved and no one gene is necessary or sufficient to cause the disorder, the statistical power of this method is vastly reduced. For example, if there were five genes involved in a disorder, and having any three was sufficient to result in disorder, then there would be individuals with the disease-causing version of the first gene who had the disorder, but also some who didn't. Furthermore, in this example there are five genes to find rather than just one. More realistically, there are likely to be ten or more genes involved in most complex traits, which interact in complicated ways with one another and with the environment. This makes it much harder to trace the effect of any one gene down through a family, as the waters will be muddied by the many other factors involved. This may be part of an explanation as to why it has been difficult to identify risk alleles for complex disorders such as depression using linkage strategies.

Another difficulty with the use of linkage to study complex traits is that often the normal variation is as much of interest as any possible disorder, and as such assigning individuals in a pedigree to being disordered or not becomes impracticable. For this reason new variations on the method using simpler strategies have been devised. These still need family members, as the fundamental basis of linkage is to see that a certain stretch of DNA is preferentially inherited with the disease. The main alternative has been the affected sibling-pair design in which genes are analyzed within sibling pairs concordant or discordant for the disease. Variations on this method have been described in detail elsewhere (Spector, Snieder, & MacGregor, 1999; Fulker & Cherny, 1996), but the main principle is to sample pairs of siblings where both members have the disease or a high level of the trait of interest. The hypothesis is that within a sample of such pairs there will be excess sharing of the risk allele. For example, if you classify sibling pairs in terms of their neuroticism levels, you would expect those where *both* siblings had high neuroticism (concordant pairs) to have the same genotype for a marker thought to be related to neuroticism. In contrast, you would expect pairs where only one individual has a high neuroticism score, and the other an average or low score (discordant pairs) not to have the same genotype. In reality the data are not as clean as this, and what you are looking for is an increased level of allele-sharing in concordant pairs as compared to discordant pairs.

Association

A more powerful method for finding QTLs (genes of small effect size in complex traits) is association (Plomin, Owen, & McGuffin, 1994; Risch & Merikangas, 1996). This is a very simple method in which the genetic marker is incorporated into a traditional case-control design. The null hypothesis is that there is equal allele frequency in a group of individuals with a certain disorder as compared to that found in the control group. Any difference in the genotype frequencies between the case and control groups suggests that the gene may be related to the disorder used to identify the groups. Alternatively the groups may be defined from being at the top and bottom extreme ends of a distribution of a complex trait, but again it is simply an allele frequency difference between the groups that is indicative of having identified a gene with an effect on that trait. For repeat markers which are functional an even simpler analysis can identify a relationship between the gene and the trait—the correlation between the number of repeats and the scores on the measured phenotype.

There are two main criticisms of the association method. The first relates to type-I error, or false positives. There are hundreds of genes that have a biochemical role relevant to behavioural traits (e.g., being one aspect of a neurotransmitter pathway), which are described as "candidate genes." There are also thousands of genes for which there is, as yet, no known role. Testing such high numbers of markers necessarily results in a type-I error problem. It is of course possible to use a Bonferroni correction as in any other study, but given that we know that any individual QTL is likely to have a very small effect size, correcting any genetic analysis in this way would almost certainly result in there being no significant associations identified. This leads to 2 main suggestions for molecular genetic work. First, collaborative efforts are needed in order that very large sample sizes can be utilised. Second, replication is the clearest protector against false positives, and molecular genetic studies therefore need to move towards incorporating replication samples into their designs.

The second criticism of association has been that there may be other factors accounting for between group differences in allele frequencies—most obviously population stratification. For example if you had a case group in which there were more individuals with brown eyes than in the control group, you might conclude that a gene that was in truth related to eye color was related to the trait or disease of interest. As there are variations of this kind between different populations it is important that case and control samples are well matched for their population of origin. One method that incorporates this issue is to use a within-family design, such as the Transmission-Disequilibrium Test (TDT) (Allison, 1997; Sham, 1997). In this method you need both biological parents of the case individual (child). The two alleles making up the child's genotype are inherited from the parents–one allele from each parent–and are called the transmitted alleles. Those that are non-transmitted (i.e., the remaining allele for each parent) are used as if they were controls. As all alleles come from one family there is no possibility of the results being false positive due to ethnic stratification.

The main advantage of molecular genetics for social scientists is that individual markers can be thought of as specific variables. As such they can be included in a study and analyzed in a similar manner to other factors thought to be of causal importance. As described above, association analyses can be conducted within the context of the standard case and control design. It is clear that careful matching will be important, but at the moment the field has not been plagued by a large number of false positives–rather there have not been many results of any kind thus far. This is likely to be due to the problems of power and sample size given the likely small effect sizes of QTLs. However, where an association is found, it makes sense to attempt to replicate this finding in other samples. This is where social science researchers will come into their own, as they already have many well worked up samples with excellent phenotypic data, which will enable the testing of far more sophisticated hypotheses as compared to the most simple level of clinical case versus control.

Finally, collecting DNA has become very simple and inexpensive, as it can be obtained through the mail using a cheek swab kit (Freeman et al., 1997; Plomin & Ratter, 1998). The remainder of this paper will show how molecular genetic data can inform on each of the five genetic hypotheses described above and in more detail in the accompanying paper on behavioral genetics (Petrill, this volume).

MOLECULAR GENETIC RESULTS

Multivariate Genetic Analyses

One of the most common findings in behavioral research is the unusually high levels of comorbidity of disorders as compared to other areas of medical research, and the associated high trait covariation (Caron &

Rutter, 1991). This comorbidity and trait covariation is seen across the board in that it is a feature of almost all types of psychiatric diagnosis, many types of developmental delay, and a wide range of other aspects of development measured as continuous traits. High rates of comorbidity are found not only within the emotional disorders, for example, between anxiety and depression (Maser & Cloninger, 1990), and within the behavioral and disruptive disorders such as attention deficit hyperactivity disorder and conduct disorder (Tannock, 1998), but also between emotional and behavioral disorders (Russo & Beadle, 1994). Comorbidity is also seen between differing types of cognitive delay (Light & DeFries, 1995).

Behavioral genetic studies have, for over a decade, been attempting to establish whether this high comorbidity and the similarly high trait covariation is due to genetic influences involved in both phenotypes, environmental influences common to both, or a mixture of the two (Petrill, this volume). Results to date have indicated quite high genetic correlations across a wide range of phenotypes (Eley, 1997a). One of the most consistent findings of this kind relates to the co-occurrence of anxiety and depression both in adults (e.g., Kendler, Neale, Kessler, Heath et al., 1992; Kendler, Walters, Neale, Kessler et al., 1995), and in children (Eley & Stevenson, 1999b; Thapar & McGuffin, 1997). This work has more recently been extended to the field of molecular genetics, where markers in the serotonin system have been explored for a whole range of behaviors related to anxiety and depression. For example, a functional polymorphism in the promoter region of the serotonin transporter gene was initially reported to be associated with the triad of anxiety, depression, and neuroticism (Lesch, Bengel, Heils, Zhang Sabol et al., 1996). The short form of this polymorphism reduces transcriptional efficiency of the promoter in the serotonin transporter gene, resulting in decreased serotonin expression and uptake. Uptake of serotonin has been implicated in anxiety in humans and in animal models, and is the target of uptake-inhibiting antidepressant and anti-anxiety drugs (Lesch et al., 1996). The polymorphism in the promoter of the serotonin transporter gene accounts for about 3-4% of the variance in these phenotypes, which is probably quite a large effect for a QTL. This finding has not been consistently replicated (e.g., Ball, Hill, Freeman, Eley et al., 1997), although many of the samples used for replication purposes have had rather small sample sizes. However, the marker has continued to be found to be associated with a number of internalizing symptoms and disorders including both unipolar and bipolar depression (Collier, Stöber, Heils, Catalano et al., 1996), obsessive compulsive disorder

(Bengel, Greenberg, Cora-Locatelli, Altemus et al., 1999), and seasonal affective disorder (Rosenthal, Mazzanti, Barnett, Hardin et al. 1998).

Another aspect of multivariate research is to identify genetic risks that are not shared between two comorbid disorders. Interestingly, behavioral genetic data have indicated that panic disorder has a much lower genetic correlation with general anxiety and depression (Kendler et al., 1995). Thus the lack of association between this marker and panic disorder in a reasonable sample size (Deckert, Catalano, Heils, Di Bella et al., 1997) suggests that the role of this gene may be general to only some aspects of emotional symptoms and disorder. In this way behavioral genetic and molecular genetic data can inform us about the shared and distinct features of the etiology of correlated traits and comorbid disorders.

Longitudinal Genetic Analyses

The third set of hypotheses to be considered are those exploring the role of genes and the environment on continuity and change in behaviors and symptoms. This area has yet to be fully explored within the discipline of behavioral genetics, but there are data which suggest that for many phenotypes genetic influences are more strongly related to continuity than change (e.g., for cognitive development; Plomin, DeFries, McClearn, & McGuffin, 2001). The extension of these longitudinal hypotheses to molecular genetic work may result in a clearer understanding of the precursors to genetically influenced disorders.

An example of how molecular genetics might inform research on longitudinal issues comes from another finding in which there has been a considerable amount of interest. This was an association reported between the dopamine receptor D4 and the personality trait of novelty-seeking (Benjamin et al., 1996; Ebstein, Novick, Umansky, Priel et al., 1996). The dopamine receptor DRD4 contains a VNTR marker with a 48 base-pair repeat. Alleles with larger numbers of repeats (the most common of which is 7 repeats) were found to be more common in individuals with high levels of novelty seeking. This result also extended to within family analyses, such that the individuals with longer DRD4 alleles had significantly higher novelty-seeking scores than their siblings with the shorter DRD4 alleles, indicating that the difference is unlikely to be due to ethnic stratification. Variation in the DRD4 gene accounted for around 4% of the variance in novelty-seeking behaviors. Subsequent work provided not only lack of replication (e.g., Pogue-Geile, Ferrell, Deka, Debski et al., 1998), but also replication and extension of this result. The most interesting feature of this work has been not only the broad range of phenotypes to which the association extends, but also the life-span perspective of the data. The dopamine receptor D4 has now been associated with neonatal temperament (Ebstein, Levine, Geller, Auerbach et al., 1998), childhood hyperactivity (Smalley, Bailey, Palmer, Cantwell et al., 1998; Swanson, Sunohara, Kennedy, Regino et al., 1998), adult novelty-seeking in a variety of populations (e.g., Okuyama, Ishiguro, Nankai, Shibuya et al., 2000), and other related adult phenotypes such as pathological gambling (Perez de Castro, Ibañez, Torres, Saiz-Ruiz et al., 1997), and substance abuse (Kotler, Cohen, Segman, Gritsenko et al., 1997; Li, Xu, Deng, Cai et al., 1997).

Another longitudinal approach to the area is to explore disorders that come into effect at certain life stages. The clear example of success from this area is that of Alzheimer's disease in which a particular allele of the apolipoprotein E (APO-E) gene on chromosome 19 is associated with early-onset Alzheimer's disease (Corder, Saunders, Strittmatter, Schmechel et al., 1993). Interestingly, this gene has no association with late-onset Alzheimer's disease, but the allele quadruples the risk for early-onset Alzheimer's disease, a very large effect size for a single gene in a behavioural phenotype.

Group Differences

One of the most commonly cited reasons for the paucity of well-replicated results using a case-control design to identify associations between a genotype and a phenotype is population stratification. As described earlier, if the base-rates of a marker or gene vary between populations this can result in false positives. A more interesting angle on this is that group differences may be meaningful and useful. For example, there is now a small but growing body of evidence that heritability estimates vary for males and females for some traits, and there may even be different genes involved (Eley, Lichtenstein, & Stevenson, 1999a). Sex differences may also be examined using molecular genetic designs. Sex differences in individual genes could result either in mean differences (e.g., a gene only has an effect in one or other sex), or in individual differences (a gene contributes to variation within one sex to a great extent than the other). In particular the X-chromosome as a whole has been considered as a good candidate for sex specific effects because while women have two copies of the X-chromosome, men have only one, their other sex chromosome being the male-specific Y-chromosome. This makes men especially vulnerable to genetic mutations on the X-chromosome. There are several single-gene disorders that have been traced to genes on the X-chromosome including hemophilia, Duchenne muscular dystrophy and Fragile-X syndrome (Sudbery, 1998). Such disorders tend to have marked sex differences with women generally having only carrier status and full presentation then being limited to men.

In terms of QTLs the X-chromosome is also of interest to those studying normal and abnormal development, especially for traits where there are large sex differences such as aggression or depression. The monoamine oxidase genes (MAOA and MAOB) are on the X chromosome and have been associated with various aspects of behavior including aggression in males, bipolar disorder, and panic in females (Brunner, Nelen, Breakefield, Ropers et al., 1993; Preisig, Bellivier, Fenton, Baud et al., 2000; Ibanez, de Castro, Fernandez-Piqueras, Blanco et al., 2000), and it may be that there is some sex-specificity to the action of this gene.

Genetic Analyses of Extremes

The relationship between normal and abnormal development is one of the most central questions for all areas of psychology. For those disorders that can be seen as the end of a continuum, the study of those with normal development may be as influential as the study of those with disorder. Behavioral genetic studies of the relationship between normal and abnormal development have generally not found much evidence for different etiological influences on development for those in extreme groups as compared to the normal range (Deater-Deckard, Reiss, Hetherington, & Plomin, 1997; Eley, 1997b; Stevenson, Batten, & Cherner, 1992), although there are notable exceptions to this. For example, a recent study of verbal development in infancy found verbal delay to be significantly more heritable than variations in verbal development in the normal range (Dale, Simonoff, Bishop, Eley et al., 1998).

The continuation of this approach into molecular genetics can similarly be broadly grouped into two groups: disorders that appear to be qualitatively distinct from normal development and those which may be the end of a normal distribution of scores. The former category includes the majority of the single-gene disorders for which there is little variation in the normal range for aspects of the phenotype: those with the risk allele have the disorder, those without it do not. This would include many of the genetic causes of mental retardation such as Fragile-X. Fragile-X is caused by an expanded triplet repeat on the X-chromosome and is associated with mental retardation, language difficulties and overactivity, impulsivity and inattention (Plomin et al., 2001). The name comes from the fragility caused in the X-chromosome by the high numbers of repeats (more than 200), which result in the chromosome breaking easily during laboratory work. Normal variation in the number of repeats (6 to 54) is not related to normal variations in the traits affected by the disorder (e.g., cognitive development, activity level). Thus, variations in the genes which cause single-gene disorders tend to result in the disease being either present or absent—and the genes involved are unlikely to be related to normal variation in cognitive ability, or any of the other traits affected by the disease.

In contrast, for many measures of behavioral development and disorder there may be more continuity between the genetic influences on variations in the normal range and at the extremes. For example, the data reported above regarding the polymorphism in the promoter of the serotonin transporter gene transporter related initially to variation in the normal range for the personality trait of neuroticism (Lesch et al., 1996), but has subsequently been related using different analytical methods to the presence or absence of related disorders (Collier et al., 1996; Bengel et al., 1999; Rosenthal et al., 1998). This has wide implications for future genetic research as it suggests that a combination of strategies—those taking disorders as the phenotype, and those considering variation in the normal range—can be utilized in conjunction with one another. While the power to detect effects is increased by the use of extreme groups, those with normal development are easier to find and to collect data and DNA on.

Environmental Analysis Using Genetically Informative Designs

One of the clearest findings from behavioral genetics and the emerging world of QTL research is that genes do not act alone for the majority of traits related to social behaviors. They not only interact with one another, but also with the environment. For example, a recent study of adult women found an interaction between having a marriage-like relationship, age and heritability of depression (Heath, Eaves, & Martin, 1998). In this study it was found that having a marriage-like relationship acted as a protective factor reducing the impact of the genetic liability to depression, and this was a particularly strong effect in women over 30 years.

The best known example of a gene-environment (g-e) interaction for which the gene and the environment have both been specified is

Phenylketonuria (PKU), a form of moderate mental retardation. This disorder is caused by mutations in a single recessive gene, which result in the individual being unable to break down phenylalanine. Phenylalanine comes from foods, particularly red meats, and if it cannot be broken down its metabolic products build up and damage the developing brain. If an infant is diagnosed as having this mutation at birth and can live on a phenylalanine free diet then the brain damage is prevented–a very simple example of an environmental response to a single-gene disorder (Plomin et al., 2001).

The majority of g-e interaction results to date have been in the medical literature. Examples include an interaction reported between maternal smoking, the transforming growth factor alpha gene, and cleft palates in infants (Hwang, Beaty, Panny, Street et al., 1995; Shaw, Wasserman, Lammer, O'Malley et al., 1996); an interaction between estrogen use, APO-E, and cognitive decline (Yaffe, Haan, Byers, Tangen et al., 2000); and modification by the environment of the risks from BRCA1 and BRCA2 in breast cancer (Gayther, Pharoah, & Ponder, 1998).

For complex traits and disorders it will be easiest to search for g-e interactions once main effects of individual genes have been found and well-replicated. However, it is clear that the putative action of some genetic risks occurs only in the presence of certain environmental influences. It is likely that genes influencing inherited vulnerability factors (such as high neuroticism or impulsivity) will be found to interact with specific environmental stimuli, increasing the risk of a poor outcome. If such processes can be better understood it may be possible to predict and prevent future problems. A further implication of this area of work is that gene-treatment interactions are likely to be found such that certain sub-groups of individuals with particular genetic risks are more likely to be respond well to specific treatments.

The second way in which genetic risk and the environment can be considered in combination is in situations where the environment mediates the risk posed by a specific genotype. The study of the influence of gene-environment correlations on behavioral and emotional outcomes is in its infancy. We know that parental psychopathology, most notably depression, is associated with a wide array of behavioral and emotional symptoms in children (Hammen, Burge, & Adrian, 1991), and yet there have been few studies that have considered whether this is due to sharing of genes, sharing of environment, or both. For example, there is a strong association between somatic symptoms in children and poor health and emotional disorders in parents, but it is unclear whether this is due to children inheriting genes related to ill-health or somatiza, or learning illness-related behaviors directly from their parents (Hotopf, Carr, Mayou, Wadsworth et al., 1998). However, there is considerable evidence for relationships between several possible environmental mediators of parental psychopathology and emotional and behavioral symptoms in the child. These include insecure attachment of depressed mothers (Radke-Yarrow, Cummings, Kuczynski, & Chapman, 1985), the absence of a good confiding relationship in a mother's life (Goodyer, Wright, & Altham, 1988), and undesirable parenting, such as inconsistent or restrictive patterns of behavior (Kohlmann, Schumacher, & Streit, 1988; Krohne & Hock, 1991). Furthermore, demographic factors such as poor housing or poverty are also likely to be relevant.

One of the most notable findings from behavioral genetics over the past decade has been how many of these traditional measures of the "environment" have been found to be heritable to some extent (Plomin, 1995; Lichtenstein, Harris, Pedersen, & McClearn, 1992). A few studies have begun to test the hypothesis that the relationship between such "environmental" variables and child outcomes may be partially due to shared genetic influences, particularly with regard to parenting behaviors. These have been shown not only to be genetically influenced, but also to have genes in common with outcomes such as child depression symptoms (Pike, McGuire, Hetherington, Reiss et al., 1996).

It is likely that genes found to be associated with emotional and behavioral outcomes will also influence aspects of the "environment" that act as mediators of genetic effects. For example, a child may inherit a gene that predisposes to aggression, and also live in a household where aggressive responses to everyday difficulties is the norm. As such the child will effectively get a "double-dose" of factors predisposing to aggressive behavior, and would be at high risk for a poor outcome.

In summary, the field of molecular genetics is just beginning to provide data relevant to the study of social behavior and relationships. As the field progresses and more main effects of specific genes on outcome are found this will become increasingly exciting as we are able to move onto testing secondary hypotheses such as those relating to multivariate questions, the relationship between the normal and abnormal, development and change over time, and interaction with the environment.

REFERENCES

Allison, D.B. (1997). Transmission-disequilibrium test for quantitative traits. *American Journal of Human Genetics, 60,* 676-690.

Ball, D.M., Hill, L., Freeman, B., Eley, T.C., Strelau, J., Riemann, R., Spinath, F.M., Angleitner, A., & Plomin, R. (1997). The serotonin transporter gene and peer-rated neuroticism. *NeuroReport, 8,* 1301-1304.

Bengel, D., Greenberg, B.D., Cora-Locatelli, G., Altemus, M., Heilis, A., Li, Q., & Murphy, D.L. (1999). Association of the serotonin transporter promoter regulatory region polymorphism and obsessive-compulsive disorder. *Molecular Psychiatry, 4,* 463-466.

Benjamin, J., Li, L., Patterson, C., Greenburg, B.D., Murphy, D.L., & Hamer, D.H. (1996). Population and familial association between the D4 dopamine receptor gene and measures of novelty seeking. *Nature Genetics, 12,* 81-84.

Brunner, H.G., Nelen, M., Breakefield, X.O., Ropers, H.H., & van Oost, B.A. (1993). Abnormal behavior associated with a point mutation in the structural gene for monoamine oxidase A. *Science, 262,* 578-580.

Caron, C., & Rutter, M. (1991). Comorbidity in child psychopathology: Concepts, issues and research strategies. *Journal of Child Psychology and Psychiatry, 32,* 1063-1080.

Collier, D.A., Stöber, G., Heils, A., Catalano, M., Di Bella, D., Arranz, M.J., Murray, R.M., Vallada, H.P., Bengel, D., Muller-Reible, C.R., Roberts, G.W., Smeraldi, E., Kirov, G., Sham, P.C., & Lesch, K.P. (1996). A novel functional polymorphism within the promoter of the serotonin transporter gene: Possible role in susceptibility to affective disorders. *Molecular Psychiatry, 1,* 453-460.

Corder, E.H., Saunders, A.M., Strittmatter, W.J., Schmechel, D.E., Gaskell, P.C., Small, G.W., Roses, A.D., Haines, J.L., & Pericak Vance, M.A. (1993). Gene dose of apolipoprotein E type 4 allele and the risk of Alzheimer's disease in late onset families. *Science, 261,* 921-923.

Dale, P.S., Simonoff, E., Bishop, D.V.M., Eley, T.C., Oliver, B., Price, T.S., Purcell, S., Stevenson, J., & Plomin, R. (1998). Genetic influence on language delay in 2-year-olds. *Nature Neuroscience, 1,* 324-328.

Deater-Deckard, K., Reiss, D., Hetherington, E.M., & Plomin, R. (1997). Dimensions and disorders of adolescent adjustment: A quantitative genetic analysis of unselected samples and selected extremes. *Journal of Child Psychology and Psychiatry, 38,* 515-525.

Deckert, J., Catalano, M., Heils, A., Di Bella, D., Friess, F., Politi, E., Franke, P., Nöthen, M.M., Maier, W., Bellodi, L., & Lesch, K. (1997). Functional promoter polymorphism of the human serotonin transporter: Lack of association with panic disorder. *Psychiatric Genetics, 7,* 45-47.

Ebstein, R., Levine, J., Geller, V., Auerbach, J., Gritsenko, I., & Belmaker, R.H. (1998). Dopamine D4 receptor and serotonin transporter promoter in the determination of neonatal temperament. *Molecular Psychiatry, 3,* 238-246.

Ebstein, R.P., Novick, O., Umansky, R., Priel, B., Osher, Y., Blaine, D., Bennett, E.R., Nemanov, L., Katz, M., & Belmaker, R.H. (1996). Dopamine D_4 receptor (D_4DR)

exon III polymorphism associated with the human personality trait novelty-seeking. *Nature Genetics, 12,* 78-80.

Eley, T.C. (1997a). General genes: A new theme in developmental psychopathology. *Current Directions in Psychological Science, 6,* 90-95.

Eley, T.C. (1997b). Depressive symptoms in children and adolescents: Etiological links between normality and abnormality: A research note. *Journal of Child Psychology and Psychiatry, 38,* 861-866.

Eley, T.C., Lichtenstein, P., & Stevenson, J. (1999a). Sex differences in the aetiology of aggressive and non-aggressive antisocial behavior: Results from two twin studies. *Child Development, 70,* 155-168.

Eley, T.C., & Stevenson, J. (1999b). Using genetic analyses to clarify the distinction between depressive and anxious symptoms in children and adolescents. *Journal of Abnormal Child Psychology, 27,* 105-114.

Freeman, B., Powell, J., Ball, D.M., Hill, L., Craig, I.W., & Plomin, R. (1997). DNA by mail: An inexpensive and noninvasive method for collecting DNA samples from widely dispersed populations. *Behavior Genetics, 27,* 251-257.

Fulker, D.W., & Cherny, S.S. (1996). An improved multipoint sib-pair analysis of quantitative traits. *Behavior Genetics, 26,* 527-532.

Gayther, S.A., Pharoah, P.D., & Ponder, B.A. (1998). The genetic of inherited breast cancer. *Journal of Mammary Gland Biology & Neoplasia, 3,* 365-376.

Goodyer, I.M., Wright, C., & Altham, P.M.E. (1988). Maternal adversity and recent stressful life events in anxious and depressed children. *Journal of Child Psychology and Psychiatry, 29,* 651-667.

Hammen, C., Burge, D., & Adrian, C. (1991). Timing of mother and child depression in a longitudinal study of children at risk. *Journal of Consulting and Clinical Psychology, 59,* 341-345.

Heath, A.C., Eaves, L.J., & Martin, N.G. (1998). Interaction of marital status and genetic risk for symptoms of depression. *Twin Research,* 119-122.

Hotopf, M., Carr, S., Mayou, R., Wadsworth, M., & Wessely, S. (1998). Why do children have chronic abdominal pain, and what happens to them when they grow up? Population based cohort study. *British Medical Journal, 316,* 1196-1200.

Hwang, S.J., Beaty, T.H., Panny, S.R., Street, N.A., Joseph, J.M., Gordon, S., McIntosh, I., & Francomano, C.A. (1995). Association study of Transforming Growth Factor Alpha Taq1 Polymorphism and Oral Clefts: Indication of Gene-Environment Interaction in a Population-based Sample of Infants with Birth Defects. *American Journal of Epidemiology, 141,* 629-636.

Ibañez, A., de Castro, I.P., Fernandez-Piqueras, J., Blanco, C., & Saiz-Ruiz, J. (2000). Pathological gambling and DNA polymorphic markers at MAO-A and MAO-B genes. *Molecular Psychiatry, 5,* 105-109.

Kendler, K.S., Neale, M.C., Kessler, R.C., Heath, A.C., & Eaves, L.J. (1992). Major depression and generalized anxiety disorder. Same genes, (partly) different environments? *Archives of General Psychiatry, 49,* 716-722.

Kendler, K.S., Walters, E.E., Neale, M.C., Kessler, R.C., Heath, A.C., & Eaves, L.J. (1995). The structure of the genetic and environmental risk factors for six major psychiatric disorders in women: Phobia, generalized anxiety disorder, panic disorder, bulimia, major depression, and alcoholism. *Archives of General Psychiatry*, *52*, 374-383.

Kohlmann, C.W., Schumacher, A., & Streit, R. (1988). Trait anxiety and parental child-rearing behaviour: Support as a moderator variable? *Anxiety Research*, *1*, 53-64.

Kotler, M., Cohen, H., Segman, R., Gritsenko, I. , Nemanov, L., Lerer, B., Kramer, I., Zer-Zion, M., Kletz, I., & Ebstein, R.P. (1997). Excess dopamine D4 receptor (*D4DR*) exon III seven repeat allele in opioid-dependent subjects. *Molecular Psychiatry*, *2*, 251-254.

Krohne, H.W., & Hock, M. (1991). Relationships between restrictive mother-child interactions and anxiety of the child. *Anxiety Research*, *4*, 109-124.

Lesch, K.P., Bengel, D., Heils, A., Zhang Sabol, S., Greenburg, B.D., Petri, S., Benjamin, J., Müller, C.R., Hamer, D.H., & Murphy, D.L. (1996). Association of anxiety-related traits with a polymorphism in the serotonin transporter gene regulatory region. *Science*, *274*, 1527-1531.

Li, T., Xu, K., Deng, H., Cai, G., Liu, J., Liu, X., Wang, R., Xiang, X., Zhao, J., Murray, R.M., Sham, P.C., & Collier, D.A. (1997). Association analysis of the dopamine D4 gene exon III VNTR and heroin abuse in Chinese subjects. *Molecular Psychiatry*, *2*, 413-416.

Lichtenstein, P., Harris, J.R., Pedersen, N.L., & McClearn, G.E. (1992). Socioeconomic status and physical health, how are they related? An empirical study based on twins reared apart and twins reared together. *Social Science and Medicine*, *36*, 441-450.

Light, J.G., & DeFries, J.C. (1995). Comorbidity of reading and mathematics disabilities: Genetic and environmental etiologies. *Journal of Learning Disabilities*, *28*, 96-106.

Maser, J.D., & Cloninger, C.R. (1990). *Comorbidity of mood and anxiety disorders*. Washington, DC: American Psychiatric Press.

McGuffin, P., Owen, M.J., O'Donovan, M.C., Thapar, A., & Gottesman, I.I. (1994). *Seminars in psychiatric genetics*. London, UK: Gaskell.

Okuyama, Y., Ishiguro, H., Nankai, M., Shibuya, H., Watanabe, A., & Arinami, T. (2000). Identification of a polymorphism in the promoter region of DRD4 associated with the human novelty seeking personality. *Molecular Psychiatry*, *5*, 64-69.

Perez de Castro, I., Ibañez, A., Torres, P., Saiz-Ruiz, J., & Fernández-Piqueras, J. (1997). Genetic association study between pathological gambling and a functional DNA polymorphism at the D4 receptor gene. *Pharmacogenetics*, *7*, 345-348.

Pike, A., McGuire, S., Hetherington, E.M., Reiss, D., & Plomin, R. (1996). Family environment and adolescent depressive symptoms and antisocial behavior: A multivariate genetic analysis. *Developmental Psychology*, *32*, 590-603.

Plomin, R. (1995). Genetics and children's experiences in the family. *Journal of Child Psychology and Psychiatry*, *36*, 33-68.

Plomin, R., DeFries, J.C., McClearn, G.E., & McGuffin, P. (2001). *Behavioral Genetics*. (4th ed.). New York: Worth Publishers.

Plomin, R., DeFries, J.C., McClearn, G.E., & Rutter, M. (1997). *Behavioral Genetics*. (3rd ed.). New York: W.H. Freeman.

Plomin, R., Owen, M.J., & McGuffin, P. (1994). The genetic basis of complex human behaviors. *Science, 264*, 1733-1739.

Plomin, R., & Rutter, M. (1998). Child development, molecular genetics, and what to do with genes once they are found. *Child Development, 69*, 1223-1242.

Pogue-Geile, M., Ferrell, R., Deka, R., Debski, T., & Manuck, S. (1998). Human novelty seeking personality traits and dopamine D4 receptor polmorphisms: A twin and genetic association study. *American Journal of Medical Genetics (Neuropsychiatric Genetics), 81*, 44-48.

Preisig, M., Bellivier, F., Fenton, B.T., Baud, P., Berney, A., Courtet, P., Hardy, Golaz, J., Leboyer, M., Mallet, J., Matthey, M.L., Mouthon, D., Neidhart, E., Nosten-Bertrand, M., Stadelmann-Dubuis, E., Guimon, J., Ferrero, F., Buresi, C., & Malafosse, A. (2000). Association between bipolar disorder and monoamine oxidase A gene polymorphisms: Results of a multicenter study. *American Journal of Psychiatry, 157*, 948-955.

Radke-Yarrow, M., Cummings, E.M., Kuczynski, L., & Chapman, M. (1985). Patterns of attachment in two- and three year olds in normal families and families with parental depression. *Child Development, 56*, 884-893.

Risch, N., & Merikangas, K.R. (1996). The future of genetic studies of complex human diseases. *Science, 273*, 1516-1517.

Rosenthal, N., Mazzanti, C., Barnett, R., Hardin, T., Turner, E., Lam, G., Ozaki, N., & Goldman, D. (1998). Role of serotonin transporter promoter repeat length polymorphism (5-HTTLPR) in seasonality and seasonal affective disorder. *Molecular Psychiatry, 3*, 175-177.

Russo, M.F., & Beidel, D.C. (1994). Comorbidity of childhood anxiety and externalizing disorders: Prevalence, associated characteristics, and validation issues. *Clinical Psychology Review, 14*, 199-221.

Sham, P.C. (1997). Transmission/disequilibrium test (TDT) for multi-allele loci. *American Journal of Human Genetics, 61*, 774-778.

Shaw, G.M., Wasserman, C.R., Lammer, E.J., O'Malley, C.D., Murray, J.C., Basart, A.M., & Tolarova, M.M. (1996). Orofacial Clefts, Parental Cigarette Smoking, and Transforming Growth Factor-Alpha Gene Variants. *American Journal of Human Genetics, 58*, 551-561.

Smalley, S.L., Bailey, J.N., Palmer, C., Cantwell, D.P., McGough, J., Del'Homme, M., Asarnow, J., Woodward, J.A., Ramsey, C., & Nelson, S. (1998). Evidence that the dopamine d4 receptor is a susceptibility gene in attention deficit hyperactivity disorder. *Molecular Psychiatry, 3*, 427-430.

Spector, T.D., Snieder, H., & MacGregor, A.J. (1999). *Advances in twin and sib-pair analysis*. London: Greenwich Medical Media.

Stevenson, J., Batten, N., & Cherner, M. (1992). Fears and fearfulness in children and adolescents: A genetic analysis of twin data. *Journal of Child Psychology and Psychiatry, 33*, 977-985.

Sudbery, P. (1998). *Human molecular genetics*. Singapore: Addison Wesley Longman Limited.

Swanson, J.M., Sunohara, G.A., Kennedy, J.L., Regino, R., Fineberg, E., Wigal, T., Lerner, M., Williams, L., LaHoste, G., & Wigal, S. (1998). Association of the dopamine receptor D4 (DRD4) gene with a refined phenotype of attention deficit hyperactivity disorder (ADHD): A family based approach. *Molecular Psychiatry, 3,* 38-41.

Tannock, R. (1998). Attention deficit hyperactivity disorder: Advances in cognitive, neurobiological, and genetic research. *Journal of Child Psychology and Psychiatry, 39,* 65-99.

Thapar, A., & McGuffin, P. (1997). Anxiety and depressive symptoms in childhood–a genetic study of comorbidity. *Journal of Child Psychology and Psychiatry, 38,* 651-656.

Yaffe, K., Haan, M., Byers, A., Tangen, C., & Kuller, L. (2000). Estrogen use, APOE, and cognitive decline: Evidence of gene-environment interaction. *Neurology, 54,* 1949-1954.

Genetic and Environmental Influences on Preschool Sibling Cooperation and Conflict: Associations with Difficult Temperament and Parenting Style

Kathryn S. Lemery
H. Hill Goldsmith

SUMMARY. With a sample of 524 pairs of three to eight year old twins recruited from a population registry, sibling cooperation and conflict were linearly related to the level of temperamental difficulty expressed in the twin dyad. Heritability accounted for 75% of the variance in Difficult Temperament, 0% of the variance in Instigating Cooperation, and 41% of the variance in Instigating Conflict. The shared environment, on the other hand, accounted for 0% of the variance in Difficult Temperament, 61% of the variance in Instigating Cooperation, and 28% of the variance in Instigating Conflict. Genetic influences largely accounted for the association between temperament and the sibling relationship. Using identical twin difference scores, Sibling Cooperation was one aspect of the nonshared environmental influence on temperament. Parental Positivity and Parental Negativity predicted the sibling relationship above and beyond temperament. Thus, the behavior genetic design eluci-

Kathryn S. Lemery is affiliated with the Arizona State University. H. Hill Goldsmith is affiliated with the University of Wisconsin-Madison.

Address correspondence to: Kathryn S. Lemery, PhD, Department of Psychology, Box 871104, Arizona State University, Tempe, AZ 85287 (E-mail: klemery@asu.edu).

[Haworth co-indexing entry note]: "Genetic and Environmental Influences on Preschool Sibling Cooperation and Conflict: Associations with Difficult Temperament and Parenting Style." Lemery, Kathryn S., and H. Hill Goldsmith. Co-published simultaneously in *Marriage & Family Review* (The Haworth Press, Inc.) Vol. 33, No. 1, 2001, pp. 77-99; and: *Gene-Environment Processes in Social Behaviors and Relationships* (ed: Kirby Deater-Deckard, and Stephen A. Petrill) The Haworth Press, Inc., 2001, pp. 77-99. Single or multiple copies of this article are available for a fee from The Haworth Document Delivery Service [1-800-HAWORTH, 9:00 a.m. - 5:00 p.m. (EST). E-mail address: getinfo@haworthpressinc.com].

dated the etiological distinctions among these correlated family variables. *[Article copies available for a fee from The Haworth Document Delivery Service: 1-800-HAWORTH. E-mail address: <getinfo@haworthpressinc.com> Website: <http://www.HaworthPress.com> © 2001 by The Haworth Press, Inc. All rights reserved.]*

KEYWORDS. Genetics, temperament, siblings

The sibling relationship influences children's social development (Dunn, 1993), and conflict between siblings can continue into adulthood (Milgram & Ross, 1982). Most children have siblings, and spend as much or more time with them than with their parents (Crouter & McHale, 1989). In the past, this within-family relationship has been under-emphasized because of the focus on between-family differences. With accumulating evidence of the importance of nonshared environmental influences on behavior (McGuire & Dunn, 1994; Rowe & Plomin, 1981), investigators are beginning to obtain measures on more than one individual per family and explore within-family differences. The present study examines the origins of differences in sibling relationships using a behavior genetic design, and considers temperamental and parenting influences on sibling cooperation and conflict.

Our beliefs about sibling relationships stem from several developmental theories. Attachment theory asserts that early relationships with attachment figures lead to internal working models that are later applied to other relationships (Bowlby, 1973; Bretherton, 1985; Sroufe & Fleeson, 1986). Similarly, social learning theory maintains that children learn behaviors in their relationships with their parents and siblings, and then apply these social skills to other relationships, such as friendships and peer relationships (MacDonald & Parke, 1984; Parke, MacDonald, Beitel & Bhavnagri, 1988; Putallaz, 1987). Cultural norms also influence how one behaves in various relationships (Stocker & Dunn, 1990). From an individual differences perspective, characteristics of the child, such as temperament, elicit similar responses from different people (Scarr & McCartney, 1983), creating similarity across different relationships. The relative importance of the processes specified by these theories has not been determined, and their importance may change across development.

The sibling relationship has been associated with other family and friendship variables. For example, sibling competition and control were associated with positive friendships (Stocker & Dunn, 1990) and insecure attachment styles (Teti & Ablard, 1989; Volling & Belsky, 1992). They were also associated with parental behaviors. Differential parental affection, responsiveness, and control toward siblings led to more conflictual and less friendly sibling relationships (Brody, Stoneman & McCoy, 1994; Dunn & Munn, 1986; Stocker, Dunn & Plomin, 1989; Volling & Belsky, 1992). Higher marital satisfaction and less spousal conflict in front of the children were also correlated with less sibling conflict (Brody, Stoneman, McCoy, & Forehand, 1992). The focus of the current study is on temperament and parenting as correlates of the sibling relationship, so we review these findings in more depth.

Temperament. There are two hypotheses as to how temperament and sibling relationship variables are related, the 'lack of fit' hypothesis and the buffering hypothesis. Munn and Dunn (1989) examined the relationship between temperament and sibling conflict with 43 sibling pairs when the second child was 24 and 36 months old. They used observational measures of sibling conflict in play interactions and mothers' ratings of temperament using the Toddler Temperament Scale or Behavior Styles Questionnaire (Fullard, McDevitt, & Carey, 1984; McDevitt & Carey, 1978). Interestingly, they found that conflict between siblings at 36 months was of greater duration in dyads in which there was a *mismatch* between the siblings in their temperamental characteristics. Specifically, they found that differences in the temperament questionnaire scales Withdrawal, Unadaptability, Mood, Persistence, and Threshold at 24 months were significantly correlated with conflict at 36 months, as were concurrent differences in Unadaptability, Intensity, and Mood. These associations suggest that temperamental differences, or lack of fit, between siblings creates an environment conducive to conflict, at least for these dimensions of temperament at these ages.

The second hypothesis, the buffering hypothesis, was suggested by Brody, Stoneman, and Burke (1987). With an older, middle childhood sample of 40 sibling pairs, they examined the association between the sibling relationship and temperament. They observed play interactions and had mothers' complete the Activity, Emotional Intensity, and Persistence scales from Martin's (1984) Temperament Assessment Battery. High Activity, high Emotional Intensity (with sisters only), and low Persistence levels were associated with increased sibling agonism. Conversely, high Persistence was associated with lower levels of agonism (with sisters and younger brothers) and more prosocial behavior (with brothers). The first implication of these findings is that the effects of temperament can be additive; that is, two active children display more agonism. The second implication is that temperament can serve a protective or buffering function (Brody et al., 1987). Positive temperamental characteristics such as high persistence in one sibling may buffer or protect the sibling relationship from agonism and conflict. This buffering hypothesis is in contrast to Dunn and Munn's lack of fit hypothesis, which would predict more conflict in dissimilar siblings.

Two additional studies partially support one or both of these hypotheses. With a sample of 67 middle childhood same-sex sibling pairs, Stoneman and Brody (1993) obtained mother and father report of Activity and Adaptability, composites formed from the Revised Dimensions of Temperament Survey (Windle & Lerner, 1986). They videotaped sibling interactions in the home during structured games, and older siblings also completed the Sibling Relationship Questionnaire (Furman & Buhrmester, 1985). They found that older sibling activity, and the interaction between older and younger sibling activity, predicted Negativity/Conflict between siblings. The age of the older sibling and the interaction between older sibling and younger sibling activity predicted Positivity/Warmth. The warmest relations were seen when siblings were similar in activity, and least warm when the older sibling was less active and younger sibling was more active. This study provided some support for the lack of fit hypothesis; that is, the sibling relationship was less warm when siblings were dissimilar in activity. However, this study also provided some support for the buffering hypothesis; that is, active younger siblings and less active older siblings had moderate conflict whereas if the older child was the one with high activity, there was higher conflict.

Stocker, Dunn, and Plomin (1989) provided some support for the buffering hypothesis with a middle childhood sample of 96 sibling pairs. Siblings participated in structured and unstructured play interactions. Using a maternal semi-structured interview to assess activity, fear, anger, sociability, emotionality, and shyness, younger siblings' activity and older siblings' frequency of emotional upset were associated with a more negative sibling relationship. More competitive sibling relationships were correlated with younger siblings' anger, intensity of emotion, and activity. On the other hand, older sibling shyness was correlated with less competitive and less controlling sibling relationships, a finding that supports the buffering hypothesis. They did not examine differences in sibling temperament to test the lack of fit hypothesis.

Other studies also considered the association between temperament and the sibling relationship but did not examine similarities and differences in sibling temperaments, so the lack of fit or buffering hypotheses cannot be supported or refuted.

Brody, Stoneman, and McCoy (1994) obtained mother and father report on Martin's (1984) Temperament Assessment Battery for 71 middle childhood same-sex sibling pairs. Parental scores were combined to yield a "difficult temperament" measure. Four years later, each sibling completed the Sibling Relationship Questionnaire (Furman & Buhrmester, 1985). Difficult temperament in the older sibling was associated with less Positivity, and difficult temperament in the younger sibling was associated with Negativity four years later.

Using the same sample, Brody, Stoneman, and Gauger (1996) found that sibling relationship quality in middle childhood was predicted by older child difficult temperament, parent-older child relationship quality (the more positive the mother-child and father-child relationship, the more positive the sibling relationship, especially if the older child's temperament was more difficult), and parent-younger child relationship quality. Thus, the parent-child relationship can be a protective factor for the effects of a difficult temperament on the sibling relationship.

In the current study, we examine parental childrearing practices and beliefs to determine whether or not parenting is related to the sibling relationship above and beyond child temperament.

Parenting Style. Parenting style was correlated with aspects of the sibling relationship in several studies. Compared to uninvolved or facilitative mothering, overcontrolling and intrusive mothering at three years predicted more conflictual and aggressive sibling interactions at six years (Volling & Belsky, 1992). Similarly, Brody et al. (1996) found that punitive and restrictive mothering was related to agonistic sibling interactions. Volling and Elins (1998) found more sibling harmony when the father disciplined the older sibling more and mother disciplined both equally. If the older sibling was disciplined by both parents more frequently, then the older child was more likely to display internalizing and externalizing symptoms. In summary, parental negativity and control have been associated with sibling conflict–but not sibling cooperation–in several studies. The sibling relationship correlates of parental positivity are less clear.

Summary. The complexity of the temperament/sibling conflict relationship is still in dispute. Whereas Munn and Dunn found that "lack of fit" between siblings creates an environment conducive to conflict with a young sample, Brody et al. concluded that a buffering effect could reduce the conflict between siblings with different temperaments with a somewhat older sample. In any case, the literature suggests a complex association, with several potential moderators, such as parenting style.

The present study utilizes a behavior-genetic (BG) design to further explore the sibling relationship. The BG method considers particular dimensions of behavior in groups that are related genetically to different degrees. Identical, or monozygotic (MZ) twins share 100% of their genes, whereas fraternal, or dizygotic (DZ) twins share on average 50% of their segregating genes. The extent to which members of twin pairs differ gives insights into the contributions of genetics and environment for that behavior. For example, if MZ twins are no more alike on a particular behavior than DZ twins, then we would conclude that the environment, rather than genetics, determines variability for that behavior. When more than one behavior is studied, the etiology of the covariance between these behaviors can be disentangled. Thus, we examine the extent to which genetic (or environmental) effects mediate the phenotypic correlation between difficult temperament and sibling cooperation and conflict.

Our BG design advances the sibling literature by addressing etiology. Additional advantages include a large sample and a relatively narrow age range (three to eight years). The age difference between siblings is also controlled for (participants are twins), which will eliminate different age gaps between siblings and minimize such confounds as parental differential treatment (e.g., Stocker et al., 1989, found that mothers treat older and younger siblings differently, which is associated with more competition between siblings).

GOALS OF THE CURRENT PROJECT

The main objective of this project is to address the following issues:

1. The literature suggests a positive relationship between various aspects of difficult temperament–emotionality, mood, and activity level–and sibling conflict. The results are less clear concerning the relationship between difficult temperament and positive aspects of the sibling relationship. In our large sample of 524 pairs of three to eight year old twins, we predict a positive correlation between difficult temperament and sibling conflict, and a negative correlation between difficult temperament

and sibling cooperation. We also explore whether or not these relationships differ by age or sex.

2. To address the lack of fit and buffering hypotheses concerning the association between temperament and the sibling relationship, we form extreme temperament groups (Difficult-Difficult, Difficult-Easy, Easy-Easy) and examine whether sibling pairs with similar or different temperaments have more or less cooperation and conflict. We hypothesize that sibling pairs that are similarly Easy will be the most cooperative and conversely, pairs that are similarly Difficult will have the most conflict, with dissimilar pairs intermediate on relationship measures. This finding would support an additive hypothesis.

3. Using the twin design, we document the magnitude of genetic and environmental influences on temperament and sibling cooperation and conflict. Do they have similar etiologies? Our measure of difficult temperament has never been subjected to a BG analysis, but we anticipate that it will yield results similar to other dimensions of negative temperament, strong genetic influences with no influence of the shared environment. Measures of the sibling relationship have never been subjected to a BG design with this age group. We hypothesize that sibling cooperation and conflict will be less genetically influenced than difficult temperament.

4. Further, we use bivariate genetic models to examine whether temperament and the sibling relationship have the same or different genetic and environmental influences, with no a priori hypotheses.

5. To consider contextual influences independent of genetic influences, we examine associations with differences between identical twins. Are differences in temperament associated with cooperation or conflict, implying that the sibling relationship is one aspect of the nonshared environmental influence on temperament? Without controlling for genetic influences, Munn and Dunn (1989) found that sibling conflict was associated with temperamental differences.

6. Last, we examine whether child rearing practices and beliefs are related to the sibling relationship and child temperament. We hypothesize that parental negativity and control are associated with sibling conflict above and beyond child temperament. We have no a priori hypotheses concerning parental positivity.

METHOD

Participants

The twin sample consisted of 524 early childhood twin pairs recruited from the Wisconsin Twin Panel, a population-based panel of all twins born in the state of Wisconsin since 1989. Participation was voluntary. The mean age was 5.06 years, ranging from 3.06 to 8.56 years. The ethnic makeup of the twins was 96% Caucasian, 2% Hispanic, 1% African-American, .5% American-Indian, and .5% Asian-American. At the time of the birth of the twins, the educational breakdown for the twins' mothers follows: 1.9% did not graduate from high school, 40.3% were high school graduates, 26.2% had one to three years of college, 20.7% were college graduates, and 11.0% had some graduate education.

Zygosity was determined by a phone interview with the primary caregiver, and a picture of the twins was sent when necessary. There were 203 MZ and 317 DZ (160 same-sex) twin pairs. Three pairs had ambiguous zygosity and were excluded from the behavior genetic analyses.

Measures

The *Zygosity Questionnaire for Young Twins* (Goldsmith, 1991). Although diagnosing zygosity by bloodtyping or "DNA fingerprinting" are the preferred methods, they are also expensive and can be difficult to justify for young children. The zygosity questionnaire for young twins yields over 95% agreement with bloodtyping and is a practical alternative.

The *Preschool Characteristics Questionnaire* (PCQ; Finegan, Niccols, Zacher & Hood, 1989) is a measure of difficult temperament in three- to six-year-olds. Children with a difficult temperament are characterized by low adaptability to novelty and high moodiness. The PCQ has been used in the past to predict later internalizing and externalizing behavioral problems. It is an upward extension of the Child Characteristics Questionnaire (CCQ; Lee & Bates, 1985). Specifically, 11 of the 32 CCQ items were modified slightly for use with older children. Twenty-one items were taken directly from the CCQ. Finegan et al. (1989) report that Cronbach's alpha for the total scale was .87. They also reported a 4-factor solution (Persistent/Unstoppable, Negative Adaptation and Affect, Difficult, and Irregular) that accounted for 36.1% of the variance.

For the present study, the mean of all 32 items was taken as our measure of difficult temperament. Alpha was .84 for a sample including one twin selected randomly from each pair, and also .84 for a replication sample including the other twin from each pair. On a 7-point scale, caregivers rate how their child compares to a typical child of the same age. An example item is, "How easy or difficult is it for you to calm or soothe your child when he/she is upset?"

Sibling Relationship Scale (SRS; Hembree, 1996). The SRS assesses the quality of young siblings' relationships (Vandell & Bailey, 1992). It has 32 items that assess jealousy, aggression, prosocial behaviors, and sibling companionship, similar to Furman and Buhrmester's (1985) Sibling Relationship Questionnaire for use with older children. For the purposes of the present study, this scale was modified for twins (i.e., 'older sibling' was replaced with 'Twin A,' and 'younger sibling' was replaced with 'Twin B'). All items were counterbalanced such that the same questions were asked about both twins. Caregivers indicate, on a 4-point scale from 'not at all true' to 'very true,' how well the twins get along. An example item is, "The twins' squabbles usually end with physical fights."

The SRS includes two factors of 16 items each: Sibling Cooperation, which is the mean of all positive items, and Sibling Conflict, which is the mean of all negative items. Alphas from the present study were .85 and .87, respectively. Twin specific subscales were also formed with items that tapped instigating or initiating sibling cooperation or conflict within the dyad. "Twin A minds when Twin B plays with his/her things" is an example item on Twin A's Instigating Conflict subscale. Instigating Cooperation contains four items, and Instigating Conflict contains five items. Items that do not tap instigation such as, "The twins share their toys with each other" were not included in these subscales. Using one twin selected randomly from each pair (sample 1) and the cotwins in a replication sample (sample 2), the average alphas for the subscales were .62 and .68, respectively.

Child Rearing Practices Report (CRPR; Block, 1965). The CRPR includes items tapping child-rearing attitudes, values, behaviors, and goals. Kochanska, Kuczynski, and Radke-Yarrow (1989) have shown that scores on the CRPR directly relate to parental behavior in interactions with their child in the laboratory. We modified the original Q-sort format in which parents arrange each item on a 7-point scale from "most descriptive" to "least descriptive." The modified CRPR asks caregivers in a questionnaire format about their attitudes on raising children using a 6-point scale from 'strongly disagree' to 'strongly agree.' Example items include, "I believe that children should be seen and not heard," and "I encourage my twins to be independent of me."

We performed a principal-components analysis with varimax rotation on the 14 scales of the CRPR. We obtained a two-factor solution that accounted for 55% of the variance. The following scales loaded on the first factor, Parental Positivity: Openness to Expression, Expression of Affect, Rational Guiding of the Child, Encourage Independence, Supervision of the Child, and Investment. Likewise, the following scales loaded on the second factor, Parental Negativity: Authoritarian Control, Control through Anxiety Induction, Control through Guilt Induction, Emphasis on Early Training, and Negative Affect. Each scale was unit weighted when forming the factor scores to encourage future replication. Three scales of the CRPR did not load on either of the two factors: Emphasis on Achievement, Protective, and Nonpunishment.

The Negativity factor describes childrearing practices that are characterized by a high degree of control, using guilt and anxiety induction, and endorsing strict rules and requirements. The Positivity factor indicates the willingness of the parent to share feelings and experiences with their children and to show affection, acceptance, and responsiveness to the children's needs.

Procedure

The twins' primary caregiver (92% mothers, 6% fathers, and 2% other caregiver) was interviewed by phone. During the forty-five minute interview, the primary caregiver responded to questions regarding the twins' zygosity and health status, temperament, sibling relationship, and their own child-rearing practices and beliefs.

Accounting for Sex and Age Effects

The differences in the scale means between girls and boys were analyzed using independent sample t-tests. There were no sex differences for Difficult Temperament. Boys were higher than girls on Sibling Conflict, $t(526) = -3.20$, $p < .01$, and also on Instigating Conflict, $t(526) = -2.03$, $p < .05$ for sample 1, and $t(526) = -2.91$, $p < .01$ for sample 2. There were no sex differences for Sibling Cooperation, but girls were higher than boys on Instigating Cooperation, $t(526) = 2.63$, $p < .01$ for sample 1, and $t(526) = 2.01$, $p < .05$ for sample 2. To explore sex differences at the dyadic level, we considered the differences among girl-girl, boy-boy, and boy-girl DZ twins. There were no differences for Difficult

Temperament, Sibling Conflict, or Instigating Conflict. Girl-girl DZ twin pairs were significantly higher on Sibling Cooperation, F (2, 316) = 6.43, $p < .01$, and Instigating Cooperation, F (2, 316) = 6.02, $p < .01$, than boy-boy or boy-girl pairs.

Difficult temperament was negatively correlated with age ($r = -.19$, $p < .01$ for sample 1, and $r = -.14$, $p < .01$ for sample 2). Similarly, Sibling Conflict was negatively associated with age, $r = -.12$, $p < .01$. Sibling Cooperation, Instigating Conflict, and Instigating Cooperation were not significantly correlated with age. These results were consistent with the literature. Because our focus was not on sex and age effects (and we have a truncated age span), we regressed out the effects of sex and linear and quadratic effects of age on Difficult temperament and all of the sibling relationship scales. It is important to do this with behavior genetic analyses, because sex and age can inflate estimates of twin similarity (McGue & Bouchard, 1984).

Behavior Genetic Statistical Approach

We used the software Mx (Neale, 1994) to fit biometric models. We started with the full, univariate model that evaluates the effects of additive genetic influences (A), common environmental influences (C), and nonshared environmental influences (E). If the estimate for C was nonsignificant, we also tested for nonadditive (dominant) genetic influences (D). D and C may not be estimated at the same time. The full models are called the univariate ACE and ADE models. We then dropped parameters to fit simpler, more specific models. Similarly, we used bivariate genetic models to decompose the covariation between measures in addition to the variation of the measures themselves. Again, we started with the full model, ACE-ACE, and dropped parameters to fit reduced models.

If the model holds and is identified, then the fit is represented in large samples as chi square with degrees of freedom equal to the number of independent values in the covariance matrix minus the number of unknowns being estimated. The chi square test is used to conclude that a model does not fit the data, so a small chi square corresponds to good fit and a large chi square corresponds to bad fit. The chi square difference test was used to compare simpler models to the full model and determine which nested model fit best. A nonsignificant difference in the chi square values between two models implies that the additional specification did not significantly reduce the fit; thus, the new, more restricted model is tentatively accepted as the more parsimonious model.

Nonnested models are compared using Akaike's Information Criterion (AIC). A small AIC is associated with better fit.

RESULTS

We first explored the phenotypic associations between the sibling relationship and temperament, including an extreme group analysis. Next we used the twin method to decompose the variance in temperament and the directional sibling relationship variables into their genetic and environmental components. We then considered the genetic and environmental etiology of the association between temperament and the sibling relationship. Using MZ difference scores on temperament, we investigated effects of context independent of genetic influences. Last, we explored specificity in the relationships with the child rearing variables, Parental Positivity and Parental Negativity.

On individual difference scales (i.e., Difficult Temperament, Instigating Cooperation, Instigating Conflict), all phenotypic analyses were conducted with a sample composed of one random twin from each pair (sample 1), and replicated with another sample consisting of the remaining twin from each pair (sample 2).

Phenotypic Associations Between Temperament and the Sibling Relationship

We hypothesized that PCQ difficult temperament would be negatively associated with sibling cooperation and positively associated with sibling conflict. Table 1 illustrates the phenotypic correlations between temperament and the sibling relationship variables separately for samples 1 and 2. Results were consistent with our hypothesis, for both the total sibling cooperation and conflict scales, and also the instigating cooperation and conflict scales that contained only the directional items (see Method).

The correlation between Sibling Cooperation and Sibling Conflict was $-.27$ ($p < .01$), which was reduced but still significant for Instigating Cooperation and Instigating Conflict ($-.18$ for sample 1, $-.17$ for sample 2, $p < .01$). Although the sibling cooperation and conflict variables were negatively correlated, they tapped largely separate aspects of the sibling relationship.

Mean Differences on Sibling Cooperation and Conflict Among Temperament Extreme Groups

As reviewed in the Introduction, the literature supports two somewhat competing hypotheses concerning similarity between co-twins on temperament, and the extent to which the dyad experiences conflict. To address this issue, we identified subgroups along the difficult temperament dimension for an extreme group contrast. First, we selected all twin pairs in which both twins scored greater than 1/2 standard deviation above the mean. These we termed the Difficult-Difficult group. Next, we selected all twin pairs in which one twin scored greater than 1/2 standard deviation above the mean, and the other twin scored less than 1/2 standard deviation below the mean, the Easy-Difficult group. Finally, we selected all twin pairs in which both twins scored less than 1/2 standard deviation below the mean, the Easy-Easy group. Table 2 gives the samples sizes and means for each of these extreme groups on the standardized, sex and age-adjusted Sibling Cooperation and Sibling Conflict scales.

Higher sibling cooperation was found in the Easy-Easy group, with the Easy-Difficult group falling near the mean, and the Difficult-Difficult group displaying the least cooperation, with linear trend analysis $F(1,213) = 42.35$, $p < .001$. Conversely, the Difficult-Difficult group displayed the most sibling conflict, the Difficult-Easy group was near the mean, and the Easy-Easy group displayed the least amount of conflict, $F(1,213) = 87.93$, $p < .001$. Our data supported an additive hypothesis, such that lack of difficult temperament in the dyad was

TABLE 1. Associations of Difficult Temperament and Sibling Conflict

| | Difficult Temperament | |
	for one twin/pair	for the other twin/pair
Sibling Cooperation	−.27	−.31
Sibling Conflict	.37	.39
Instigating Cooperation	−.26	−.29
Instigating Conflict	.37	.41

N = 524 individuals in each sample, all correlations are significant at p < .01.

TABLE 2. Mean Differences Among Extreme Temperament Groups on Sibling Cooperation and Conflict

Temperament Pairing in the Family	Number of Pairs	Mean (SD) Cooperation	Mean (SD) Conflict
Difficult/Difficult	78	−0.14 (.43)	0.39 (.51)
Easy/Difficult	37	−0.09 (.48)	−0.04 (.42)
Easy/Easy	99	0.23 (.30)	−0.29 (.47)

Temperamental classification based on scores .5 standard deviation above (Difficult) or below (Easy) the mean on PCQ Difficult Temperament. Scores are standardized, sex, age, and age² adjusted. SD = standard deviation.

associated with sibling cooperation, and extreme difficult temperament in both twins was associated with the most sibling conflict. Our finding also supported the buffering hypothesis because if one twin had an Easy temperament, the dyad was buffered from extreme conflict. However, we found no support for the lack of fit hypothesis, which would predict more conflict in dissimilar pairs, for this particular measure of temperament.

Using the twin method, we next decomposed the variance in temperament and instigating sibling cooperation and conflict into their genetic and environmental components using the full sample.

Genetic and Environmental Influences on Temperament and the Sibling Relationship

Difficult Temperament. The intraclass correlations for PCQ Difficult Temperament are presented in the top portion of Table 3. Overall, MZ correlations were higher than both same-sex and opposite-sex DZ correlations, indicating a genetic effect on Difficult Temperament. We next estimated the genetic and environmental influences on temperament using biometrical model fitting, with same-sex and opposite-sex DZ twins in separate groups. We first fit an additive ACE model (see Method for description). C was nonsignificant and thus it was dropped from the model. Next, we tested for effects of genetic nonadditivity, using an ADE model. However, Difficult temperament was best represented by an AE model; standardized estimates and model fit indices are given in Table 3. Thus, difficult temperament was highly heritable and genetic influences entirely accounted for twin similarity. This pat-

TABLE 3. Twin Intraclass Correlations and Estimates from the Most Parsimonious Biometric Model

	MZ	Same-sex DZ	Opposite-sex DZ	h^2	c^2	e^2	χ^2	df	p	AIC
Difficult Temperament	.76	.21	.37	.75	.—	.25	7.14	7	.41	−6.86
Instigating Cooperation	.64	.57	.60	.—	.61	.39	8.86	7	.26	−5.14
Instigating Conflict	.68	.55	.42	.41	.28	.31	5.54	6	.48	−6.46

N = 515 preschool twin pairs (200 MZ pairs, 159 same-sex DZ and 156 opposite-sex DZ pairs). All correlations were significant at p < .01. Standardized estimates: h^2 = heritability, c^2 = shared (common) environment, and e^2 = nonshared environment. AIC = Akaike's Information Criterion.

tern of results is typical for parent report of negative dimensions of temperament (see review in Goldsmith, Buss & Lemery, 1997).

Sibling Cooperation and Conflict. We computed twin concordance for Sibling Cooperation and Sibling Conflict. In contrast to Difficult Temperament, there was a substantial influence of the shared environment on both of these scales (see Table 3). Using model fitting, we found that the most parsimonious model for Sibling Cooperation was the CE model; thus, twin similarity could be entirely accounted for by the shared environment. For Sibling Conflict, the most parsimonious model was the ACE model, with both genetic and shared environmental influences on twin similarity. These findings indicate that temperament and the sibling relationship have somewhat different etiologies, with temperament being more influenced by genes, and the sibling relationship being more influenced by the shared environment.

Biometrically Decomposing the Relationship Between Temperament and the Sibling Relationship

Next, we used bivariate model fitting to decompose the phenotypic association between temperament and sibling cooperation and conflict. To what extent did the individual difference variable of temperament share genetic and/or environmental influences with the interpersonal sibling relationship variables? The bottom of Table 1 gives the phenotypic correlations that we decomposed.

We started with the full ACE-ACE bivariate model and systematically dropped parameters to test reduced, nested models (e.g., AE-AE). The most parsimonious model for Difficult Temperament and Instigating Cooperation was the AE-CE model displayed in Figure 1. This model fit the data well, c^2 (25) = 26.92, p = 0.36, AIC = −23.08. This model suggested that the small genetic influence on Instigating Cooperation was entirely shared with temperament. Also, this genetic influence entirely accounted for the association between the variables; there were no environmental influences in common.

FIGURE 1. Bivariate model of difficult temperament and instigating sibling cooperation. The latent variable A represents additive genetic influences, C represents shared (common) environmental influences, and E represents nonshared environmental influences including measurement error. The subscript c represents influences common to both temperament and Instigating Sibling Cooperation, whereas the subscript u represents influences unique to Instigating Sibling Cooperation. The figure depicts estimates for one twin. A is correlated 1.00 for MZ twins (who share all their genes) and .50 for DZ twins (who share on average 50% of their segregating genes). C is correlated 1.00 for both MZ and DZ twins reared together, and E is not correlated for either MZ or DZ twins. Difficult Temp = PCQ Difficult Temperament; Sibling Coop = Instigating Cooperation. Fit statistics: χ^2 = 26.92, df = 25, p = 0.36, AIC = −23.08.

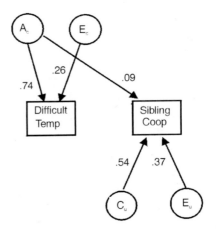

We next considered the association between Difficult Temperament and Instigating Conflict. The most parsimonious model was the AE-ACE, illustrated in Figure 2. This model fit the data well, χ^2 (21) = 23.08, p = 0.34, AIC = -18.92. Instigating Conflict shared some genetic etiology with temperament, but also had a unique genetic component. The association between the variables was decomposed into largely genetic, but also some environmental, influences.

Thus, shared genetic influences, although small, largely accounted for the association between temperament and sibling cooperation and conflict using bivariate genetic model fitting.

Differences in MZ Twins

One way to consider contextual influences, independent of genetic influences, in a twin design is to investigate the differences between MZ twins on a particular trait. These associations do not hold for DZ twin intrapair differences, where there are genetic sources of difference. In order to examine how the sibling relationship may impact the twins, we examined the association between Sibling Cooperation and Conflict

FIGURE 2. Bivariate model of difficult temperament and instigating sibling conflict. Difficult Temp = PCQ Difficult Temperament; Sibling Conflict = Instigating Conflict. Fit statistics: χ^2 = 23.09; df = 23, p = 0.46, AIC = -22.91.

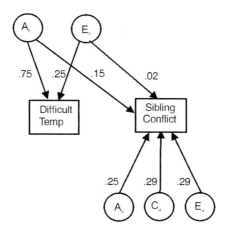

and absolute MZ intrapair differences in Difficult Temperament. Sibling Cooperation significantly and negatively predicted MZ differences in Difficult Temperament, $r = -.24, p < .01$. On the other hand, the correlation between Sibling Conflict and MZ differences in temperament did not reach significance, $r = .13, p = .06$. Therefore, Sibling Cooperation was one aspect of the environment that was associated with differences in identical twins' temperaments.

Associations with Child Rearing Practices

Table 4 portrays the correlations among the parenting factors, Parental Positivity and Parental Negativity (see Methods for details of the principal components analysis), and the temperament and sibling relationship variables. Interestingly, both parenting variables were unrelated to Difficult Temperament. Although the parenting factors were orthogonal, they both were significantly correlated with Sibling Cooperation. This pattern was not true for Sibling Conflict. Parental Negativity, but not Parental Positivity, was significantly positively correlated with Sibling Conflict. These results suggest that Sibling Cooperation and Sibling Conflict have different correlates at the phenotypic level. Additionally, parenting was not associated with temperament, yet it was associated with the sibling relationship variables.

We used linear regression to determine whether or not parenting contributed to the prediction of the sibling relationship above and beyond child temperament. We entered twin one temperament, twin two temperament and the interaction in the first block, then Parental Negativity and Parental Positivity in the second block. A stepwise selection algorithm was used with a criterion of $p = .05$ for entry and $p = .10$ for removal. When

TABLE 4. Associations of Temperament and Sibling Cooperation and Conflict with Parenting Style

	Parental Positivity	Parental Negativity
Difficult Temperament	−.07	.11˙
Sibling Cooperation	.16*	−.18*
Sibling Conflict	.01	.26*

* $p < .01$, ˙ $p = .015$, N = 524, r's with Difficult Temperament are sample 1 and 2 averages.

predicting Sibling Cooperation, R-squared for temperament was .12 ($p <$ .001), changes in R-squared for Parental Positivity and Parental Negativity were .02 ($p < .001$) and .04 ($p < .001$), respectively. Similarly when predicting Sibling Conflict, R-squared for temperament was .20 ($p < .001$), and the change in R-squared for Parental Negativity was .04 ($p < .001$). Parental Positivity was not a significant predictor of Sibling Conflict. In conclusion, parenting predicted the sibling relationship above and beyond child temperament.

DISCUSSION

The sibling relationship is part of a dynamic family system. Positive and negative aspects of this relationship are correlated with both individual difference measures, such as temperament, and other relationship measures such as attachment style and best friendships. Investigators are beginning to take process-oriented approaches to understanding these associations. For example, Brody et al. (1996) found that temperament moderated the association between parent-child relationship quality and sibling relationship quality. The present study also took a process-oriented approach—using a behavior genetic design to examine the etiology of these associations.

With a sample of 524 pairs of three to eight year old twins, we found that difficult temperament was associated with less cooperation and more conflict in the sibling relationship, which replicated Brody et al.'s (1994) study with an older sample. Munn and Dunn (1989) found that siblings who were *dissimilar* in temperament had more conflict. Additionally, Brody and colleagues (1987) found that persistence buffered the sibling relationship from conflict. For difficult temperament, we hypothesized an additive influence, with siblings that were similarly difficult displaying the most conflict and least cooperation, and those who were similarly easy displaying the least conflict and most cooperation, with dissimilar pairs intermediate on both scales.

Using extreme temperamental groups, our data supported the additive hypothesis. Because the dissimilar Easy-Difficult pairs displayed less conflict than the Difficult-Difficult pairs, Easy temperament in one twin acted as a buffer on the sibling relationship—similar to persistence in Brody et al. (1987). Summarizing across studies, temperamental difficulty, emotional intensity and activity level have been linked to negative sibling relationships, whereas easiness, persistence and shyness protect or buffer the relationship. Future studies could advance this literature by determining whether or not other temperamental characteristics, such as interest and pleasure, also buffer the relationship. Additionally, future studies could systematically consider sibling differences for a variety of dimensions of temperament to elucidate whether or not these differences contribute to conflict for some dimensions. For Difficult Temperament in the present study, they did not.

With genetic model fitting, we estimated that heritability accounted for 75% of the variance in Difficult Temperament, 0% of the variance in Instigating Cooperation, and 41% of the variance in Instigating Conflict. The shared environment, on the other hand, accounted for 0% of the variance in Difficult Temperament, 61% of the variance in Instigating Cooperation, and 28% of the variance in Instigating Conflict. Genetic influences largely accounted for the association between Difficult temperament and Instigating Cooperation and Conflict. We also documented a relationship between Sibling Cooperation and MZ differences in Difficult Temperament. This finding indicates that the sibling relationship is one aspect of a nonshared environmental process involving Difficult Temperament. Thus, the behavior genetic design has distinguished the etiological influences among these correlated individual and family variables.

Several studies in the literature suggested that overcontrolling, punitive, and restrictive parenting was linked to more agonism, conflict and aggression in the sibling relationship (Volling & Belsky, 1992; Brody et al., 1996). We replicated this finding with a significant correlation between Parental Negativity and Sibling Conflict. Parental Positivity was unrelated to Sibling Conflict but was positively associated with Sibling Cooperation. The parenting factors were unrelated to temperament, and predicted the sibling relationship above and beyond child temperament. Perhaps parenting was one aspect of the large shared environmental influence on the sibling relationship measures.

The primary limitation of the current study is that the sibling relationship, temperament, and parenting data stem from the same source, the twins' primary caregiver. However, there are several patterns in our data that suggest a general bias such as parent personality did not account for the observed associations. First, we found quite different patterns of biometric findings. Fraternal twins were not very similar on Difficult Temperament ($R = .21$), but quite similar on Instigating Cooperation ($R = .57$) and Instigating Conflict ($R = .55$), for example. Also, the parenting measures were unrelated to temperament, but were associated with the sibling relationship, illustrating differential external validity.

Furthermore, parent report measures of the sibling relationship, difficult temperament, and parenting have been found to be valid measures that relate to observational measures. Parent and self report of positive and negative aspects of the sibling relationship using Furman and Buhrmester's (1985) Sibling Relationship Questionnaire were related to lab measures of the same constructs (e.g., Stoneman & Brody, 1993, for self report). Likewise, parent report measures of temperament were related to lab measures of the same constructs (Eaton, 1994; Kochanska, Coy, Tjebkes, & Husarek, 1998). Similarly, scores on the CRPR directly relate to parental behavior in interactions with their child in the laboratory (Kochanska, Kuczynski, & Radke-Yarrow, 1989).

This study simultaneously took into account genetic and psychological factors when examining the sibling relationship. This process-oriented approach advances the sibling relationship literature. Obvious future directions would be to expand the design longitudinally and include multiple measures of the sibling relationship, temperament and parenting. We are currently pursuing these expansions.

REFERENCES

Block, J. H. (1965). *The child rearing practices report.* Berkeley, CA: Institute of Human Development, University of California.

Bowlby, J. (1973). *Attachment and loss, vol 2: Separation, anxiety and anger.* New York: Basic Books.

Bretherton, I. (1985). Attachment theory: Retrospect and prospect. In I. Bretherton & E. Waters (Eds.), Growing points of attachment theory and research, pp. 3-35. *Monographs of the Society for Research in Child Development, Serial No. 209, 50* (1-2).

Brody, G. H., Stoneman, Z., & Burke, M. (1987). Child temperaments, maternal differential behavior, and sibling relationships. *Developmental Psychology, 23,* 354-362.

Brody, G. H., Stoneman, Z., & Gauger, K. (1996). Parent-child relationships, family problem-solving behavior, and sibling relationship quality: The moderating role of sibling temperaments. *Child Development, 67,* 1289-1300.

Brody, G. H., Stoneman, Z., & McCoy, J. K. (1994). Forecasting sibling relationships in early adolescence from child temperaments and family processes in middle childhood. *Child Development, 65,* 771-784.

Brody, G. H., Stoneman, Z., McCoy, J. K., & Forehand, R. (1992). Contemporaneous and longitudinal associations of sibling conflict with family relationship assessments and family discussions about sibling problems. *Child Development, 63,* 391-400.

Crouter, A. & McHale, S. (1989). *Childrearing in dual- and single-earner families: Implications for the development of school-age children.* Paper presented at the Biennial Meeting of the Society for Research in Child Development, Kansas City, MO.

Dunn, J. (1993). *Young children's close relationships: Beyond attachment.* London: Sage.

Dunn, J., & Munn, P. (1986). Sibling quarrels and maternal intervention: Individual differences in understanding and aggression. *Journal of Child Psychology and Psychiatry, 27,* 583-595.

Eaton, W. O. (1994). Temperament, development, and the Five-Factor Model: Lessons from activity level. In C. F. Halverson, Jr., G. A. Kohnstamm, & R. P. Martin (Eds.), The developing structure of temperament and personality from infancy to adulthood (pp. 173-187). Hillsdale, NJ: Erlbaum.

Finegan, J., Niccols, A., Zacher, J., & Hood, J. (1989). Factor structure of the preschool characteristics questionnaire. *Infant Behavior and Development, 12,* 221-227.

Fullard, W., McDevitt, S. C., & Carey, W. B. (1984). Assessing temperament in one to three-year-old children. *Journal of Pediatric Psychology, 9,* 205-217.

Furman, W. & Buhrmester, D. (1985). Children's perceptions of the qualities of sibling relationships. *Child Development, 56,* 448-461.

Goldsmith, H. H. (1991). A zygosity questionnaire for young twins: A research note. *Behavior Genetics, 21,* 257-269.

Goldsmith, H. H., Buss, K. A., & Lemery, K. S. (1997). Toddler and childhood temperament: Expanded content, stronger genetic evidence, new evidence for the importance of environment. *Developmental Psychology, 33,* 891-905.

Hembree, S. (1996). *Parental contributions to young children's sibling relationships.* Unpublished doctoral dissertation, University of Wisconsin, Madison, WI.

Kochanska, G., Coy, K. C., Tjebkes, T. L. & Husarek, S. J. (1998). Individual differences in emotionality in infancy. *Child Development, 69,* 375-390.

Kochanska, G., Kuczynski, L., & Radke-Yarrow, M. (1989). Correspondence between mothers' self-reported and observed child-rearing practices. *Child Development, 60,* 56-63.

Lee, C. L., & Bates, J. E. (1985). Mother-child interaction at age 2 years and perceived difficult temperament. *Child Development, 56,* 1314-1325.

MacDonald, K. & Parke, R. E. (1984). Bridging the gap: Parent-child play interaction and peer interactive competence. *Child Development, 55,* 1265-1277.

Martin, R. P. (1984). *Manual for the temperament assessment battery.* Unpublished monograph, University of Georgia, Athens.

McDevitt, S.C., & Carey, W.B. (1978). The measurement of temperament in 3-7 year old children. *Journal of Child Psychology and Psychiatry, 19,* 245-253.

McGue, M. & Bouchard, T. J. Jr. (1984). Adjustment of twin data for the effects of age and sex. *Behavior Genetics, 14,* 325-343.

McGuire, S., & Dunn, J. (1994). Nonshared environment in middle childhood. In J. C. DeFries, R. Plomin, & D. W. Fulker (Eds.), *Nature and nurture during middle childhood.* Cambridge, MA: Blackwell.

Milgram, J. I. & Ross, H. G. (1982). Effects of fame in adult sibling relationships. *Individual Psychology: Journal of Adlerian Theory, Research and Practice, 38,* 72-79.

Munn, P., & Dunn, J. (1989). Temperament and the developing relationship between siblings. *International Journal of Behavioral Development, 12,* 433-451.

Neale, M. C. (1994). *Mx: Statistical modeling* (2nd ed.). Box 710 MCV, Richmond, VA 23298: Department of Psychiatry, Medical College of Virginia.

Parke, R., MacDonald, K., Beitel, A., & Bhavnagri, N. (1988). The role of the family in the development of peer relationships. In K. Kreppner & R. M. Lerner (Eds.), *Family systems and life span development.* Hillsdale, NJ: Erlbaum.

Putallaz, M. (1987). Maternal behavior and children's sociometric status. *Child Development, 54,* 417-426.

Rowe, D. C., & Plomin, R. (1981). The importance of nonshared (E₁) environmental influences in behavioral development. *Developmental Psychology, 17,* 517-531.

Scarr, S., & McCartney, K. (1983). How people make their own environments: A theory of genotype-environment effects. *Child Development, 54,* 424-435.

Stocker, C., & Dunn, J. (1990). Sibling relationships in childhood: Links with friendships and peer relationships. *British Journal of Developmental Psychology, 8,* 227-244.

Stocker, C., Dunn, J., & Plomin, R. (1989). Sibling relationships: Links with child temperament, maternal behavior, and family structure. *Child Development, 60,* 715-727.

Stoneman, Z., & Brody, G. H. (1993). Sibling temperaments, conflict, warmth, and role asymmetry. *Child Development, 64,* 1786-1800.

Sroufe, L. A., & Fleeson, J. (1986). Attachment and the construction of relationships. In W. Hartup & Z. Rubin (Eds.), *Relationships and development* (pp. 51-71). New York: Cambridge University Press.

Teti, D. M., & Ablard, K. E. (1989). Security of attachment and infant-sibling relationships: A laboratory study. *Child Development, 60,* 1519-1528.

Vandell, D. L., & Bailey, M. D. (1992). Conflicts between siblings. In C. Shantz & W. W. Hartup (Eds.), *Conflict in Child and Adolescent Development.* Cambridge, UK: Cambridge University Press.

Volling, B. L., & Belsky, J. (1992). The contribution of mother-child and father-child relationships to the quality of sibling interaction: A longitudinal study. *Child Development, 63,* 1209-1222.

Volling, B. L., & Elins, J. L. (1998). Family relationships and children's emotional adjustment as correlates of maternal and paternal differential treatment: A replication with toddler and preschool siblings. *Child Development, 69,* 1640-1656.

Windle, M. & Lerner, R. M. (1986). Reassessing the dimensions of temperamental individuality across the life span: The Revised Dimensions of Temperament Survey (DOTS-R). *Journal of Adolescent Research, 1,* 213-230.

Family Process as Mediator
of Biology × Environment Interaction

Kristin Riggins-Caspers
Remi J. Cadoret

SUMMARY. The present study used Baron and Kenny's (1986) mediated moderation model to explore a potential mediator–parenting behaviors–of an interaction between biological risk and adoptive parent psychopathology which has been shown to significantly predict adolescent problem behaviors in a sample of adult adoptees (M_{age} = 25; n = 133). The outcomes of interest were retrospective reports on adoptee adolescent nonaggressive conduct disordered behaviors and aggres-

Kristin Riggins-Caspers and Remi J. Cadoret are affiliated with the University of Iowa.

The authors wish to acknowledge the cooperation of the following individuals and agencies in making this study possible: James Yeast, Catholic Charities of the Archdiocese of Dubuque; Craig Mosher, PhD, and Nancylee Ziese, Hillcrest Family Services; Paul Loeffelholz, MD, Iowa Department of Corrections, Des Moines; Marge Corkery, Iowa Department of Human Services; and Leonard Larson, PhD, Lyn Lienhard, and Louise Koch, Lutheran Social Service of Iowa. The authors thank the following individuals who performed data collection for this study: Michelle Appel, Alberta Conrad, Nancy Copeland, Donna Davis, Rebecca Hansel, Roxane Moller, Carol Moss, Scott Murray, Helen Peck, Chris Richards, Jody Speer, Sommai Ung, and Angela Wright.

Data collection for this project was supported by Grant R01 DA05821 from the National Institute on Drug Abuse awarded to the second author. Writing of the manuscript was supported by a National Research Scientist Award (T32MH14620) presented to the first author.

Address correspondence to: Kristin Riggins-Caspers, Psychiatry Research/MEB, University of Iowa, Iowa City, IA 42242 (E-mail: kristin-caspers@uiowa.edu).

[Haworth co-indexing entry note]: "Family Process as Mediator of Biology × Environment Interaction." Riggins-Caspers, Kristin, and Remi J. Cadoret. Co-published simultaneously in *Marriage & Family Review* (The Haworth Press, Inc.) Vol. 33, No. 2/3, 2001, pp. 101-130; and: *Gene-Environment Processes in Social Behaviors and Relationships* (ed: Kirby Deater-Deckard, and Stephen A. Petrill) The Haworth Press, Inc., 2001, pp. 101-130. Single or multiple copies of this article are available for a fee from The Haworth Document Delivery Service [1-800-HAWORTH, 9:00 a.m. - 5:00 p.m. (EST). E-mail address: getinfo@haworthpressinc.com].

101

sive/oppositional behaviors. Predictors were biological risk in the birth parents of the adoptees, psychopathology in the adoptive parents, and retrospectively reported parenting behaviors of the adoptive parents. The mediated moderation model was not supported due to the nonsignificance of the biological risk × adoptive parent psychopathology interaction term. However, support for an independence model was supported in that biological risk interacted with maternal warmth and overprotection to predict adolescent adoptee behavior. Greater maternal warmth decreased problem behaviors among adoptees with a biological risk for psychopathology; whereas greater maternal overprotection increased adolescent problem behaviors among adoptees with biological risk. Fathers' parenting behaviors did not interact with biological risk to predict adolescent adoptee behavior. Finally, the findings were consistent across type of behavior suggesting a homogenous effect. *[Article copies available for a fee from The Haworth Document Delivery Service: 1-800-HAWORTH. E-mail address: <getinfo@haworthpressinc.com> Website: <http://www. HaworthPress. com> © 2001 by The Haworth Press, Inc. All rights reserved.]*

KEYWORDS. Biology-environment interaction, externalizing behavior problems, discipline

Over the past few decades, our research group has utilized the adoption paradigm to examine the role of biology and environment in the manifestation of psychopathology (Cadoret & Cain, 1981; Cadoret, Cain, & Crowe, 1983; Cadoret, Yates, Troughton, Woodworth et al., 1995; Cadoret, Winokur, Langbehn, Troughton et al., 1996; Cutrona, Cadoret, Suhr, Richards et al., 1994). The adoption paradigm has allowed us to explore the independent contributions, as well as the interactive effects, of biology and environment to psychopathology.[1] The testing of biology × environment interaction is both empirically and clinically important in that it can provide information on the malleability of biological diatheses and identify environmental factors that can be targeted for intervention. In this paper, we extend our previous research by examining specific family processes that might be involved in the attenuation or augmentation of biological risk for psychopathology; specifically, adolescent symptoms of aggressivity and conduct disorder.

Biology × environment interaction refers to the dependence of the phenotypic expression of a biologically influenced trait on the environment in which it is expressed (Plomin, DeFries, & Loehlin, 1977). In order to test for biology × environment interaction, specific biological and environmental factors must be identified and proven to be independent such that there is no evidence of selective placement or evocative biology-environment correlation.[2] In our studies, an adoption paradigm was used to separate biological from environmental effects and to examine their independent, as well as synergistic effects, on adolescent and adult psychiatric diagnoses.

Alcoholism and antisocial personality in the biological parents of our adopted-away offspring served as indicators of a biological risk for psychopathology and any psychiatric disorder in the adoptive parents served as an indicator of additional environmental risk. In earlier studies we found that biological and environmental risk made independent contributions to psychopathology in the adoptee with the presence of psychopathology in either the biological parent or the adoptive parent predicting higher rates of problem behaviors (Cadoret et al., 1995, 1996). More interesting, however, has been the emergence of evidence for biology × environment interaction between our indicators of biological and environmental risk. Specifically, adoptees with both a biological parent and an adoptive parent with a diagnosable psychiatric disorder showed significantly higher rates of psychopathology than adoptees without any risk factors or adoptees having only a single risk factor–biological or environmental.

Although the above findings are interesting, the environmental risk factor, adoptive parent psychopathology, included in the biology × environment interaction term could be classified as a distal environmental factor (Bronfenbrenner, 1977, 1986; Bronfenbrenner & Ceci, 1994). Distal environmental effects are those factors that affect an individual indirectly through some mediating process, whereas proximal environmental effects are those factors that have a direct impact on an individual. For example, economic stress–a distal environmental effect–has been found to negatively affect adolescent outcomes because economically stressed parents demonstrate fewer supportive behaviors towards family members–a proximal environmental effect (Conger, Conger, Elder, Lorenz et al., 1992; Conger, Lorenz, Elder, Melby et al., 1991; Elder & Caspi, 1988; Larzalere & Patterson, 1990; Lempers, Clark-Lempers, & Simons, 1989).

The environmental factor in our previous work was psychopathology in the adoptive parent, which is clearly external to the adopted child leaving identification of the proximal environmental factor responsible for mediating the combined effects biological and environmental risk on psychopathology relatively unexplored in these data. Therefore, we have identified a proximal environmental factor, parenting behaviors that we believe will account for the observed effects of the biology × environment interaction on adoptee adolescent behavior.

The model tested in this paper was derived primarily from research examining the influence of biology on the quality of the parent-child relationship (Plomin, McClearn, Pederson, Nesselroade, & Bergeman, 1988; Rowe, 1981, 1983). Specifically, researchers have used the concept of evocative biology-environment correlation to examine the effects of heritable behavioral characteristics to the quality on parenting received. Evocative biology-environment correlation is demonstrated when a characteristic shown to have a heritable component elicits a specific response from the environment (Plomin et al., 1977; Scarr & McCartney, 1983). Our research group and others have demonstrated that parenting behaviors can in fact be affected by characteristics of the child and that biological contributions to this evocative process are possible (Anderson, Lytton, & Romney, 1986; Bell, 1968, 1979; Bell & Chapman, 1986; Caspi, Elder, & Bem, 1987; Dodge, 1990; Lytton, 1982, 1990; Maccoby, 1992; Patterson, 1982, 1983; Reid, Patterson, & Loeber, 1982). Initially, the biological effects were nonspecific and derived from twin analyses (Rowe, 1981, 1983). However, more recent investigations utilizing specific measures of the environment have confirmed the effect of heritable characteristics on certain dimensions of the environment (Ge, Conger, Cadoret, Neiderhiser et al., 1996; O'Connor, Deater-Deckard, Fulker, Rutter et al., 1998; Riggins-Caspers, Cadoret, Knutson, & Langbehn, 2000).

Somewhat neglected has been the impact of adoptive parent characteristics on the above evocative process (Riggins-Caspers et al., 2000). It is now well accepted that psychiatric symptoms, such as depression, can have a negative impact on parenting behavior (Campbell, Pierce, Moore, Marakovitz et al., 1996; Coyne & Downey, 1991; Downey & Coyne, 1990; Kendler, Sham, & MacLean, 1997; Maccoby & Martin, 1983; Susman, Trickett, Iannotti, Hollenbeck et al., 1985). Perhaps such characteristics influence the reactivity of the parent to the child's heritable behavior resulting in lowered positive parenting behaviors and increased negative parenting behaviors among parents of children at risk for psychopathology (Greenwald, Bank, Reid, & Knutson, 1997). By using a genetically informative design, we can test the relation between the processes of biology × environment interaction and evocative biology-environment correlation by examining whether psychiatric disorder in the adoptive parents moderates the evocative effects of biological psychopathology on the parenting behaviors of the adoptive parents (Rutter 1997a, b; Rutter & Pickles, 1991; Rutter, Dunn, Plomin, Simonoff et al., 1997).

To summarize, the present paper is an extension of previous research in that we were attempting to identify proximal family processes, such as parenting behaviors, that can help explain the role of the interaction between biological risk for psychopathology and a distal environmental risk factor–adoptive parent psychopathology–in the prediction of psychopathology. We are proposing that two specific parenting behaviors–warmth and overprotection––will mediate the above effects. The identified parenting behaviors were selected due to their confirmed association with the outcomes of interest (e.g., aggressive/oppositional problem behaviors, conduct-disordered behavior) and the potential biological influence on these parenting behaviors (Deater-Deckard, 2000; Kendler, 1996; Patterson, DeBaryshe, & Ramsey, 1989; Rey & Plapp, 1989; Rothbaum & Weisz, 1994; Stormshak, Bierman, McMahon, & Lengua, 2000). A final goal of the study was to examine whether the identified processes–biology $\times$ environment interaction and biology-environment correlation–are equally predictive of both adoptee behavioral phenotypes (e.g., nonaggressive conduct-disordered behaviors and aggressive/oppositional behavior). The findings in the literature are inconsistent with regard to the homogeneity of both biological and environmental predictors of these behaviors (Deater-Deckard & Plomin, 1999; Langbehn, Cadoret, Yates, Troughton et al., 1998; Rey & Plapp, 1990; Stormshak et al., 2000).

METHODS

Sample

Subjects from this study were recruited from four adoption agencies in the state of Iowa: Lutheran Social Services of Iowa; Catholic Charities of the Archdiocese of Dubuque; Hillcrest Family Services; and the Iowa Department of Human Services. Adult adoptees and their adoptive parents were selected on the basis of whether the adoptee had a biologic parent with specific psychiatric or behavioral problems (see Measures Section). Adoptees were further selected to be between 18 and 45 years of age ($M = 25$). A sample of adoptees without any psychiatric disorder or behavioral problems in the biological parents were matched to the case adoptees by sex and age of the adoptee, as well as age of the biological mother at the time of birth. Data was collected between the years 1989 and 1992. Twenty-five of the original 197 subjects interviewed were excluded due to placement in the adoptive home after six months of age. Twenty-two of the adoptive parents were not interviewed and these cases were also excluded from analyses. Finally, cases in which missing data were present for adoptee reports on either the adoptive mother or father were excluded (n = 17) leaving an effective sample size of 133.

The sample was predominantly Caucasian (94%) and approximately evenly split between female and male adoptees (n = 77 and n = 56). Adoptive mothers and fathers averaged one to two years of college education (M = 13 to 14 years). The median adoptive family income fell between $30,000 and $39,999 at the time of interview. The mean ages of the adoptive mothers and fathers at the time of placement were 29 years (SD = 4.09; Min, max = 20, 42) and 32 years (SD = 4.70; Min, max = 23, 55). Biological mothers and fathers averaged less than a high school education at the time of the adolescent's placement (M = 11 years) and were, on average, 21 years (SD = 4.99; Min, max = 14, 44) and 23 years (SD = 5.86; Min, max = 16, 53) of age at the time of placement.

Procedure

The data collection procedure was typical of an adoptee study in which the starting point of sample identification is the birth parent. For this study, birth parent psychopathology was the selection criterion and was established by circulating names of birth parents, who had placed infants up for adoption at birth, through the following institutions: State of Iowa Mental Health institutions, University of Iowa Hospitals and Clinics, and the Iowa Department of Corrections (see Cadoret et al., 1995). The adoption agencies were provided the names of the adoptees that were identified for inclusion in the study by the researchers at which time the adoption agencies submitted the names of the biological parents to the above agencies.

Initially, 11,700 birth parent names were screened. Identification of the correct biologic parent was confirmed by comparing the birth dates in the records with the birth date of the birth parent, place of parent's birth, names of the biologic parents, names of siblings, and other identifying data found in the institution's record with information in the adoption agency records. Using DSM-III-R criteria, three psychiatrists independently diagnosed the birth parent as antisocial, alcoholic or a substance abuser. Of the original sample of birth parents, 238 fit the criteria from which 197 adult adoptees and their families were successfully recruited (Cadoret et al., 1995). The kappa values for diagnosing antisocial personality and alcoholism were adequate (range: κ = .67-.90). Differences in diagnoses were resolved through conference, with unresolved differences in diagnosis resulting in the case being discarded.

Measures

The adoptive parents were administered a face-to-face interview designed to assess the parent's retrospective view of the adoptee's physical health, temperament, development, school achievement, and social adjustment from infancy to the present. A section of the interview also covered details about emotional or psychiatric problems in the adoptive family. All adoptive parent interviews were conducted in person by interviewers blind with respect to the biological background of the adoptee.

During a separate interview session, the adoptees were administered the Diagnostic Interview Schedule (DIS; Robins, Helzer, Cottler, & Goldring, 1989). As with the adoptive parent interviews, the interviewers were blind to the biological background of the adoptee at the time of interview. The DIS allows lay interviewers to assess current and lifetime prevalence of psychiatric illness by DSM-III-R (American Psychiatric Association, 1987) criteria and Research Diagnostic Criteria, as well as Feighner criteria (Feighner, Robins, Guze, Woodruff et al., 1972). Prior to data collection, the research assistants responsible for the interviewing attended a weeklong DIS instruction course offered by Washington University in St. Louis, Missouri. Medical records for the adoptees were also collected to verify treatment and diagnosis of psychiatric conditions or substance abuse. Prior to the interview session, adoptees also completed several questionnaires assessing retrospective recall of the adoptive family atmosphere and the parent-adolescent relationship during the time in which the adoptee resided in the adoptive home.

Environmental Risk. As part of an extensive interview about the adopted adolescent and the adoptive family, adoptive parents reported the presence or absence of psychopathology (e.g., depression, anxiety, alcoholism, other psychiatric conditions) and legal or marital problems experienced by one or both of the adoptive parents. An environmental risk variable was created by counting the number of adverse parent factors present in either the adoptive mother or the adoptive father. The count score was dichotomized into two environmental risk groups: lower environmental risk if fewer than two environmental risk factors were present (n = 75) and higher environmental risk if two or more factors were present (n = 58). The rates of individual diagnoses for adoptive parents in the lower environmental risk group ranged from 0 percent for drug problems to 16 percent for anxiety disorder. For the higher environmental risk group, the rates of individual diagnoses ranged from 5 percent for drug problems to 71 percent for depression problems (M = 46 percent).

Biological Risk. In contrast to earlier studies using this sample of adoptees in which single diagnoses served as indicators of birth parent psychiatric disorder (Cadoret et al., 1995, 1996), the presence of one of three possible diagnoses (e.g., alcoholism, antisocial personality, and criminal behavior) was used to classify an adoptee as a case (n = 77). Adoptees without any record of psychiatric or behavioral problems in the biological parents were classified as controls (n = 56). This approach was used due to the high co-morbidity of alcoholism, antisocial personality, and criminality in our sample.

Chi-square analyses failed to show significant differences across environmental risk for individual diagnoses of birth parents. Independent t-tests also failed to show significant mean differences in the number of birth parent psychiatric diagnoses between the lower risk environmental group (*M* = .83, *SE* = .13) and the higher risk environmental group (*M* = .97, *SE* = .15). These analyses suggest that adoptees from both environmental risk groups had similar biological liability for these psychiatric disorders and that selective placement based on these characteristics was absent.

Nonaggressive Adolescent Conduct Disordered Behaviors. A summation of 11 questions consistent with conduct-disordered behaviors, excluding those items describing aggressive behaviors, were summed to create a count score of nonaggressive conduct-disordered behavior (Deater-Deckard & Plomin, 1999; Rey & Plapp, 1990; Stormshak et al., 2000). Both the adoptee and the adoptive parents were asked to provide retrospective reports on the frequency (e.g., 0 = never, 1 = rarely, 2 = sometimes, 3 = often) or occurrence of each of the following behaviors during high school: vandalism, stealing, setting fires, truancy, expulsion, arrested, running away, frequent lying, engaged in early sex, poor grades, and trouble in school. A behavior was counted as absent if both the adoptive parent and the adoptee reported that the adoptee did not engage in the behavior at all or rarely. The behavior was counted present if either the adoptee or the adoptive parent reported that the adoptee engaged in the behavior at least some of the time. Internal consistency of the combined score was very good (Cronbach's α = .82).

Adolescent Aggressive/Oppositional Behaviors. Adoptive parents also retrospectively reported on the presence of adolescent aggressive/oppositional behaviors. Ten items relating to adoptee aggressive/oppositional behaviors (e.g., temper tantrums, bullying other kids, defying or not minding authority, fighting with other kids, frequent quarrelling, rebellious behavior, teasing other kids, insolence, physically abusive towards others, verbally abusive towards others) were rated as either present or absent (Deater-Deckard & Plomin, 1999; Loney, Langhorne, Paternite, Whaley-Klahn et al., 1980; Rey & Plapp, 1990; Stormshak et al., 2000). A total adolescent aggression/opposition score was computed by summing all items. The scale showed good internal consistency (Cronbach's α = .74).

Parenting Behaviors. The Parental Bonding Questionnaire (Parker, Tupling, & Brown, 1979) was used to assess parental warmth and overprotection experienced during childhood and adolescence. The adoptee rated on a 4-point scale the degree to which the parent engaged in 25 different behaviors. Examples of parental warmth are, "Spoke to me with a warm and friendly voice," "Was affectionate to me," and "Enjoyed talking things over with me." Examples of parental overprotection are, "Tried to control everything I did," "Felt I could not look after myself unless she was around," and "Invaded my privacy." The adoptee completed the PBI for their mother and father.

Analyses

The model tested was Baron and Kenny's (1986) mediated moderation model (see Figure 1). The Mediated Moderation model can be contrasted with an Independent Effects model in which interactions between biological risk and both parenting behaviors and distal environmental factors are identified as having independent, direct effects on adolescent problem behaviors (see Figure 1). With the Mediated Moderation model, parenting behaviors are identified as a potential mechanism through which the interaction between biological risk and distal environmental risk impact adolescent problem behaviors (see Figure 1). In other words, the Mediated Moderation model hypothesizes that parenting behaviors differ as a function of combined biological and distal environmental risk, whereas the Independents Effects model hypothesizes that parenting behaviors are unaffected by the interaction

FIGURE 1. Mediational and Independent Effects Models

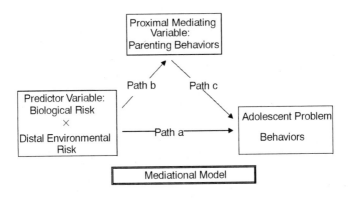

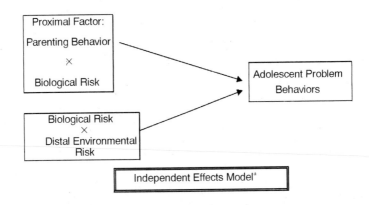

[a]The main effects for adoptee gender, parenting behaviors, environmental risk, and biological risk were included in the ANOVA models but have been omitted from the figures to simplify presentation.

between biological and distal environmental risk, and instead, interact directly with biological risk to affect behavior.

Testing for mediated moderation involves a three-step process outlined by Baron and Kenny (1986). The first step tests the significance of the association between the biology × distal environment interaction and the outcome—adolescent problem behaviors (see path a, Figure 1). The second step tests for a significant association between the biology ×

distal environment interaction and the mediating variable–parenting behaviors (see path b, Figure 1). Finally, the strength of the path tested in Step 1 is examined after both the biology × distal environment interaction term and the proximal environmental effect are entered into the model. If a significant reduction in the path coefficient (path a, Figure 1) is observed, then mediation is supported.

In order to test for mediation, the following analyses were conducted. Depending on the outcome variable (adolescent problem behaviors, parenting behaviors) either univariate analyses of variance or repeated measures analysis of variance were used. Univariate analyses of variance were used to analyze parenting behaviors because each dimension was analyzed separately as a main effect with the corresponding behavior as a covariate. A repeated measures approach was used for adolescent problem behaviors to test for homogeneity of effects across different types of behavior problems (e.g., nonaggressive conduct disordered vs. aggressive/oppositional behavior). First, a mixed repeated measures analysis of variance was used to examine the first requirement of the mediational analyses. In addition to their main effects, the interaction between biological risk and distal environmental risk were entered as a predictor of adolescent problem behaviors (see Figure 1). Step 2 of the analyses was tested by univariate analyses of variance with parenting behaviors as the dependent variable. The model was identical to that estimated in Step 1 with the main effects for biological and distal environmental risk, and their interaction, as the predictor variables. Finally, a mixed repeated measures analysis of variance predicting adolescent problem behaviors was proposed to test Step 3. The model predicting adolescent problem behaviors consisted of the following dependent factors: the main effects of the components of the biological × distal environment interaction, the biology × distal environment interaction, and parenting behaviors. Because of the repeated measures design, scores for adolescent nonaggressive conduct-disordered behaviors and aggressive/oppositional behaviors were standardized with the mean of the total sample as the standard. The scores were log-transformed prior to standardization.

RESULTS

Mediation of Biology × Environment Interaction

The following main effects were included in the mixed multivariate model used to test Step 1 of the mediation analyses: type, gender, distal environmental risk, and biological risk. The interaction between biological and distal environmental risk was also entered in the model. Type of adolescent problem behaviors was a within-subjects factor and tested for homogeneity of outcomes across type of problem behaviors (e.g., nonaggressive conduct disordered vs. aggressive/oppositional behavior) when entered as an interaction term with the other factors in the model. The goal of these analyses was to demonstrate a significant biology × environment interaction for adolescent problem behaviors. The model entered each term hierarchically using the Type I sums of squares option available in SPSS 9.0. Due to small numbers, the interactions of the between-subjects factors were limited to the biological risk × distal environment risk interaction. A summary of the significant findings for the mediation analyses is presented in Table 1.

Biology × Distal Environment Interaction: Adolescent Problem Behaviors. A significant multivariate effect was found for the interaction between behavioral type (i.e., aggressive/oppositional behavior versus non-aggressive conduct disorder) and adoptee gender, Wilks's Lambda = .96; $df = 1, 128$; $p < .05$. According to Cohen's classification of effect sizes, eta-squared (η^2) for this interaction was small (see Table 1). Follow-up analyses of the interaction showed greater gender differences for nonaggressive conduct disordered behavior, $t (131) = -4.23$, $p < .001$, than for aggressive/oppositional behavior, $t (131) = -2.07, p < .05$. For both behaviors, however, females had fewer symptoms than males (Conduct disorder: $M \pm SD = -.29 \pm .10$ versus $M \pm SD = .40 \pm .14$; Aggressive/oppositional: $M \pm SD = -.16 \pm .10$ versus $M \pm SD = .21 \pm .15$, respectively). Analyses examining the effect of type separately for males and females were nonsignificant for both suggesting that the source of the interaction was the differential gender effect (i.e., higher symptoms among males than females) as a function of type of adolescent problem behavior.

With regard to the between-subject effects, a significant multivariate effect was found for adoptee gender, $F (1, 128) = 13.45$, $p < .001$. Recall, however, that the gender main effect was qualified by its interaction with problem behavior type. A significant main effect for biological risk was found, $F (1, 128) = 8.94$, $p < .01$; however, the biology × distal environment interaction was also significant for adoptee problem behaviors, $F (1, 128) = 4.32, p < .05$. The pattern of the biology × distal environment interaction is presented in Figure 2 and shows the marginal means for the standardized problem behaviors estimated from the multivariate model. Adoptees having both biological and distal envi-

TABLE 1. Summary of Significant Findings for Mixed Repeated-Measures MANOVA

	Biological Risk × Distal Environmental Risk	Other Significant Effects
Step 1:		
Multivariate Adolescent Problem Behaviors	$\eta^2 = .03^*$	Type x Gender: $\eta^2 = .03^*$ Gender: $\eta^2 = .08^{***}$ Biological Risk: $\eta^2 = .07^{**}$
Step 2:		
Maternal Caring	n.s.	Environmental Risk: $\eta^2 = .06^{**}$
Maternal Overprotection	n.s.	n.s.
Paternal Caring	n.s.	Gender: $\eta^2 = .04^*$ Environmental Risk: $\eta^2 = .03^*$
Paternal Overprotection	n.s.	n.s.

$^*p < .05.\ ^{**}p < .01.\ ^{***}p < .001.$

ronmental risk factors (bar d) demonstrated significantly high rates of problem behaviors than adoptees without both risk factors (bar a) and adoptees with only the distal environmental risk (bar c). The effect sizes for the above findings ranged from small to medium (see Table 1). It should be noted that none of the above between-subject effects differed by behavioral type suggesting homogeneity of effects across the two types of behavior problems. The findings from Step 1 of the analyses fulfilled the first requirement of testing mediation.

Biology × Environment Interaction: Parenting Behaviors. Because the biology × distal environment interaction term was significant for problem behaviors, univariate analyses of variance were used in the second step of the analyses which tested for biology × distal environment interaction as a predictor of parenting behaviors–the proximal environmental factor (Step 2, see Figure 1). Because of documented differences in the degree of biological influence on parental warmth and overprotection, as well as differences for maternal versus paternal parenting behaviors, analyses were conducted separately by parenting behavior (caring versus overprotection) and by source of parenting (mother versus father) (Kendler, 1996; Kendler et al., 1997; Plomin & Bergeman, 1991; Rowe, 1981, 1983). The predictors in the models were adoptee gender, distal environmental risk, and biological risk. Interactions were again limited to the biology by distal environmental risk.[3]

FIGURE 2. Interaction Between Distal Environmental Risk and Biological Risk for Multivariate Adolescent Problem Behaviors

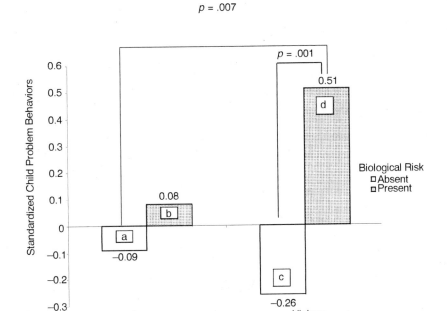

For maternal caring behaviors, the main effect for environmental risk was significant, $F (1, 128) = 8.00$, $p < .01$, and showed a medium effect size (see Table 1). Adoptees with greater distal environmental risk reported fewer caring behaviors from mother ($M \pm SD = 29.75$ (0.73 versus $M \pm SD = 26.07$ (1.14). No significant effects were found for maternal overprotection. For fathers' caring behaviors, a small effect was found for gender, $F (1, 128) = 5.67$, $p < .05$, with females reporting higher caring behaviors from fathers than males ($M \pm SD = 28.31$ (0.84 versus $M \pm SD = 25.66$ (0.94). A significant main effect for distal environmental risk was also found, $F (1, 128) = 4.37$, $p < .05$, with lower caring behaviors reported by adoptees with greater risk ($M \pm SD = 28.40$ (0.75 versus $M \pm SD = 25.43$ (1.07). Again, the effect size for this latter term was small (see Table 1). Finally, no significant effects were found for fathers' overprotection. The absence of biology × distal environment interaction for any parenting behaviors in Step 2 fails to support the second requirement of mediation; consequently, no further testing of mediation was conducted.

Independent Effects Model

Although mediation was not supported, we remained interested in examining whether our proximal environmental factors–parental warmth and overprotection–made significant contributions to problem behaviors both independently and in combination with biological factors above and beyond those contributions made by our distal environmental factor–adoptive parent psychopathology. Therefore, instead of entering parenting behavior only as a main effect in the model–which would be the case if we tested for mediation–we also included it as an environmental effect in a second biology × environment interaction term (see Figure 1, Independent Effects Model). We then evaluated whether the interaction between biological risk and the distal environmental factor–environmental risk–remained significant after controlling for the interaction between biological risk and the proximal environmental factor–adoptive parenting behaviors. If the biology × distal environment interaction term was no longer significant after entering the biology × proximal environment interaction term, then it can be proposed that the interaction between the proximal environmental risk factor and the biological risk factor was the more salient process affecting adoptees' behavior.

A mixed repeated measures MANOVA was conducted separately for caring and overprotective behaviors and for source of parenting behaviors (e.g., father versus mother). For example, maternal overprotection was entered as a covariate in the model including maternal caring behaviors as a main effect. In order to incorporate parenting behaviors into the model, a median split was used to create two groups either low or high on each parenting behavior. When one parenting dimension was included as a main effect (e.g., caring behaviors), the remaining parenting dimension was entered as a covariate (e.g., overprotection). A total of four mixed repeated measures MANOVAs were conducted. We recognize that the median split is not optimal for dichotomizing a continuous variable; however, our sample size precluded the use of more extreme cutoffs (Maxwell & Delaney, 1993). The between-subject factors were entered in the following order: adoptee gender, distal environmental risk, parenting behavior, and biological risk. Adoptee gender was included to control for gender differences in perceived parenting and overall problem behaviors. The within-subject factor was again "type" (e.g., nonaggressive conduct disordered and aggressive/ oppositional behaviors) and tested differences in mean levels of non-aggressive conduct-disordered behavior and aggressive/oppositional behaviors, as well as differences in the significance of main effects and interactions in predicting these two outcomes. Interactions among the between-subjects factors were limited to all two-way interactions between distal environmental risk, parenting behaviors, and biological risk. Higher-order interactions were not examined due to inadequate sample size. In order to examine the unique contributions of each term in the model, the Type III sums of squares option was used for these analyses. In order to reduce repetition, the discussion of significant effects will be limited to the main effects (e.g., main effects for parenting) and interactions unique to these analyses. A summary of all significant findings is presented in Table 2. For all analyses, the within-subjects factor–type of adolescent problem behavior–was nonsignificant.

   Maternal Caring. As demonstrated by the large effect size, the covariate maternal overprotection was highly significant, $F(1, 124) = 15.09$, $p < .001$, and showed a strong positive correlation with both nonaggressive conduct-disordered and aggressive/oppositional behaviors (see Tables 2 and 3). The main effects for adoptee gender and biological risk were also significant with both having a medium effect size (see Table 2). Finally, the interactions between biological risk and maternal caring behavior and between biological risk and distal environmental risk were significant, $F(1, 124) = 4.87$, $p < .05$, and $F(1, 124) = 3.98$, $p < .05$, respectively (see Table 2). The effect sizes for the interaction terms were small. Adoptees with a biological risk for psychopathology who rated their adoptive mother low on caring behavior (see bar *b*, Figure 3) showed significantly higher rates of problem behaviors than adoptees with all other combinations of factors (see bars *a*, *c*, and *d*, Figure 3). With regard to the combined effect of biological risk and distal environmental risk on outcome, the pattern of the interaction was identical to that presented in Figure 2. Note the small impact of the proximal biology × environment interaction on the point estimate for the distal biology × environment interaction term ($\Delta = .34$) suggesting the interactive effects for both distal and proximal environmental effects as represented by maternal caring behaviors are independent in this model.

   Maternal Overprotection. For the MANOVA examining maternal overprotection as a proximal effect, the main effects for adoptee gender and biological risk remained significant (see Table 2), as was the main effect for maternal overprotection, $F(1, 124) = 11.44$, $p < .001$. A me-

TABLE 2. Summary of Significant Findings for Independent Effects Multivariate Model Predicting Adolescent Problem Behaviors

Proximal Effect in Model	Proximal Effect × Biological Risk	Distal Environmental Risk × Biological Risk	Proximal Effect × Environmental Risk	Significant Main Effects
Maternal Caring Behaviors	$\eta^2 = .04^*$	$\eta^2 = .03^*$	n.s.	Mat. Overprotection (CV): $\eta^2 = .11^{***}$ Gender[a]: $\eta^2 = .07^{**}$ Biological Risk[a]: $\eta^2 = .07^{**}$
Maternal Overprotection	$\eta^2 = .04^*$	n.s.	n.s.	Gender[a]: $\eta^2 = .07^{**}$ Maternal Overprotection: $\eta^2 = .08^{***}$ Biological Risk[a]: $\eta^2 = .08^{**}$
Paternal Caring Behaviors	n.s.	n.s.	n.s.	Pat. Overprotection (CV): $\eta^2 = .10^{***}$ Gender[a]: $\eta^2 = .10^{***}$ Biological Risk[a]: $\eta^2 = .09^{***}$
Paternal Overprotection	n.s.	n.s.	n.s.	Gender[a]: $\eta^2 = .10^{***}$ Paternal Overprotection: $\eta^2 = .06^{**}$ Biological Risk[a]: $\eta^2 = .07^{**}$

Note. CV = covariate in model. [a] Significant in previous multivariate analyses (see Table 2). $^* p < .05.$ $^{**} p < .01.$ $^{***} p < .001.$

TABLE 3. Correlations Between Parenting Variables and Adolescent Problem Behaviors

		1	2	3	4	5
1	Maternal Caring Behavior					
2	Maternal Overprotection	−.55***				
3	Paternal Caring Behavior	.47***	−.31***			
4	Paternal Overprotection	−.23**	.62***	−.51***		
5	Nonaggressive Conduct Disorder Behaviors	−.35***	.33***	−.15	.26**	
6	Aggression/Oppositional Behaviors	−.26**	.39***	−.10	.28***	.62***

* $p < .05$. ** $p < .01$. *** $p < .001$.

dium effect size was observed for the latter term. Higher rates of problem behaviors were found for adoptees that rated their mothers above the median on overprotective behaviors ($M \pm SD = -.17 \pm .09$ versus $M \pm SD = .35 \pm .12$).

With regard to interactions, the interaction between biological risk and maternal overprotection, $F(1, 124) = 4.23, p < .05$, was significant and had a small effect size (see Table 2). As shown in Figure 4, adoptees with a biological risk for psychopathology that rated their mothers high on overprotective behaviors (bar *d*) demonstrated significantly higher rates of adolescent problem behaviors than adoptees with all other combinations of factors (bars *a*, *b*, and *c*). Finally, in contrast to the previous model, the interaction between the distal environmental risk factor and biological risk was no longer significant after controlling for the preceding effects, $F(1, 124) = 2.72, p > .10$, and the point estimate was reduced substantially ($\Delta = 1.60$).

Paternal Caring Behaviors. The covariate–paternal overprotection–was significant, $F(1, 124) = 13.90, p < .001$, with a medium to large effect size (see Table 2). Higher overprotection was associated with higher rates of adolescent problem behaviors (see Table 3). The main effects for adoptee gender and biological risk remained significant with medium to large effect sizes, respectively. None of the interactions were significant.

Paternal Overprotection. The main effects for adoptee gender, biological risk, and paternal overprotection, $F(1, 124) = 4.75, p < .05$, were significant and were small to medium in effect size (see Table 2). Adolescent problem behaviors were highest when the adoptive father was

FIGURE 3. Interaction Between Maternal Caring Behaviors and Biological Risk for Multivariate Adolescent Problem Behaviors

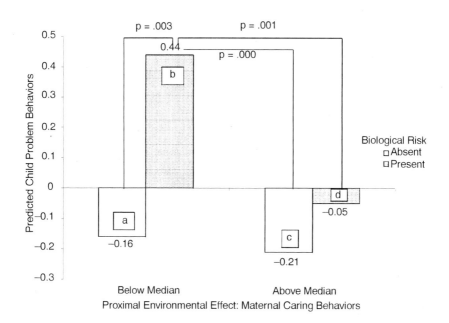

described as overprotective ($M \pm SD = -.12 \pm .10$ versus $M \pm SD = .33 \pm .13$). None of the interactions were significant.

DISCUSSIONS

The analyses in this paper used the mediated moderation model developed by Baron and Kenny (1986) to test the mediation of a significant biology $\times$ distal environment interaction by a third variable–parenting behaviors (see Figure 1). This model was proposed because earlier studies had shown a significant interaction between biological risk for psychopathology and what we have now defined as a distal environmental risk factor–psychopathology in the adoptive parent(s)–when predicting adolescent aggressive and conduct disordered behavior (Cadoret et al., 1995, 1996). By utilizing the mediated moderation

FIGURE 4. Interaction Between Maternal Overprotective Behaviors and Biological Risk for Multivariate Adolescent Problem Behaviors

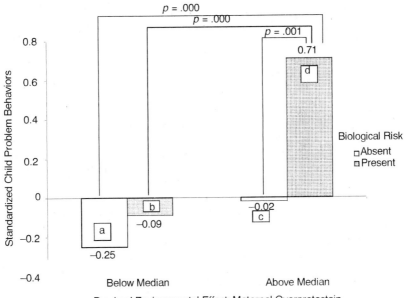

framework, we could develop a model identifying potential family processes that would account for the effect of the distal environment, in combination with biological risk, on adoptee outcomes.

We were able to partially support the mediated moderation model in that a significant interaction between biological risk (biological psychopathology) and distal environmental risk (environmental psychopathology) was found with adoptees having both risk factors demonstrating the highest rates of problem behaviors. Given the significance of the biology × distal environment interaction term predicting adolescent problem behaviors, the mediation of this effect by retrospectively reported parenting behaviors was examined. We tested whether the interaction between biological and distal environmental risk predicted maternal and paternal warmth and overprotection. None of the interactions were significant for any parenting behaviors.

The lack of statistically significant findings eliminated parenting behaviors, as we measured them, as a potential mediator of our particular biology × distal environment effect; therefore, the proposed mediated moderation model was not supported. Although not statistically signifi-cant, we examined the plots of parenting behaviors as a function of bio-logical risk and distal environmental risk and found some interesting patterns. For both maternal and paternal caring behaviors, adoptees with both biological and distal environmental risk factors rated their parents one-half of a standard deviation below the mean. All other groups were within one-tenth of a standard deviation of the mean. This pattern of scores would be consistent with an evocative effect in that parents under additional stress provided less warmth when the adoptee had a biological diathesis for psychopathology. Interestingly, the same pattern was not found for overprotective behaviors, suggesting that overprotective parenting may be more of a function of parent character-istics than a response to child characteristics (Kendler et al., 1997). However, this interpretation is subject to confirmation by future testing due to the lack of significant findings in these data.

Even though the mediated moderation model was not supported, we were still interested in exploring the role of proximal family processes in predicting adolescent problem behaviors. Therefore, we tested whether parenting behaviors, alone and in interaction with biological risk, explained unique variance in our outcomes. We found significant interactions between biological risk and both maternal caring behaviors and maternal overprotection. For maternal caring behaviors, signifi-cantly higher rates of problem behaviors were found for adoptees that had a biological risk for psychopathology and rated their parents below the median on caring behaviors. For maternal overprotection, rates of adolescent problem behaviors were highest for adoptees with biological risk that rated their mother above the median on overprotection. Fur-thermore, the previously significant interaction between biological risk and distal environmental risk was no longer significant after controlling for the interaction between biological risk and the proximal environ-mental factor. These findings demonstrate that although parenting be-haviors do not mediate the effects of the biological risk × distal environment interaction, they do contribute to the process of expressing biologically influenced traits.

Finally, we explored whether the model tested was consistent across different types of adolescent problem behaviors (e.g., nonaggressive conduct-disordered and aggressive/oppositional behaviors). The cur-rent literature is inconsistent with regard to the etiological and nosological independence of oppositional and conduct-disordered be-havior. For example, Rey and Plapp (1990) have suggested that the classification of aggressive/oppositional behavior and conduct-disordered behavior as independent psychiatric disorders should, in part, be based on the degree to which the etiology of each disorder differs. In their study, parenting behaviors were found to predict equivocally both nonaggressive conduct-disordered behaviors and aggressive/oppositional behaviors suggesting a possible continuum of behavioral severity (Rey & Plapp, 1999). Stormshak et al. (2000), on the other hand, found specific parenting practices predicted specific behavioral outcomes. Using genetic models to estimate biological and environmental components of behavior, other studies have found that the magnitude of biological and environmental contributions do differ by behavior with aggressive behaviors having a more heritable component and shared environment contributing more to delinquent behaviors (Deater-Deckard & Plomin, 1999; Eley et al., 1999). Further complicating the interpretation of the findings is the dependence of biological and environmental variation on the reporter of behaviors (Deater-Deckard & Plomin, 1999). Our findings are consistent with Rey and Plapp (1990) in that we did not find any significant interactions between our within-subjects factor–type–and any other effect in the model suggesting homogenous effects across behavior type (Rey & Plapp, 1990). However, the similarity in findings between our study and Rey and Plapp (1990) could be an artifact of methodology since both studies used the Parental Bonding Instrument, a retrospective measure of parenting, whereas those studies in which differences across predictors were found used concurrent measures of parenting and behavioral problems (Deater-Deckard, 2000; Stormshak et al., 2000).

Interpretation of Findings

There are several interesting interpretations of the findings presented above. First, examination of the significant biology $\times$ distal environment interaction shows that the quality of the distal environment can serve to suppress or augment biological diatheses for behaviors, which was evidenced by the significant biological effect in the environment with greater environmental risk. In the absence of additional environmental stressors, the biological propensity for externalizing behavior problems was not expressed. An additional example of the importance of environment in moderating biological diatheses was the interaction between biological risk and maternal caring behaviors. High rates of adolescent problem behaviors were expressed only when the adoptive mother was rated below the median on caring behaviors. The absence of a biological main effect in the high caring group suggests that maternal caring behavior can serve as a protective factor by suppressing the expression of biological diatheses for externalizing behaviors. Similar protective effects were not observed for fathers' caring behavior as demonstrated by the absence of a significant interaction between paternal caring behaviors and biological risk. The absence of an effect for father's positive parenting behaviors is interesting considering that the mean levels of caring behavior were comparable across mothers and fathers. In terms of environmental factors that increased the risk for expressing psychopathology, only maternal overprotection served as an additional proximal environmental risk factor. In conclusion, it appears that restrictive parenting approaches might serve to increase problem behaviors for children that have a propensity to be noncompliant (Bates, Pettit, Dodge, & Ridge, 1998; Volling & Belsky, 1992; Kochanska, 1997; Lengua, Wolchik, Sandler, & West, 2000) and that the effect of restrictive parenting may be parent specific with negative implications occurring only for mothers (Belsky & Volling, 1987; Belsky, Hsieh, & Crnic, 1998).

Several limitations of this study must be acknowledged. First, reports on both parenting behaviors and adolescent problem behaviors were retrospective, making the potential for recall bias and distortion possible. The issue of recall bias and interpretation of evocative effects would be considerable if a biological component to the adoptee's recall of their home environment was found. The threat to the validity of the study is that there would be no way to determine if the report on parenting was a function of actual experience, in which case evocation can be suggested, or if the reports were influenced by the tendency of persons with specific characteristics (e.g., antisocial personality) to perceive experiences in certain ways. In the latter case, biological contributions to parenting could be overestimated and evocation would be identified falsely. Therefore, researchers should attempt to acquire independent information about the experience or behavior of interest any time retrospective reports are used. In our data, we also collected adoptive parent reports on the environment using the Family Environment Scale. Correlations across measurements and across reporters showed moderate significance (min, max: $r = .20-.37$), suggesting that the adoptees' retrospective report is consistent with the reports of others within the same family. We did not include the FES in this paper because we were interested in specific behaviors directed toward the adoptee that might result from specific heritable behaviors displayed by the adoptee. Second, the range of parenting may have been limited by the use of an adoption sample, making generalization difficult and increasing the risk of null findings due to reduced variability (Stoolmiller, 1999). The fact that we did find significant results for maternal parenting behaviors suggests that a sufficient degree of variability was present. Furthermore, the equivalence of variance for both maternal and paternal caring (*SD* = 7.46 and 7.39, respectively) and overprotective behaviors (*SD* = 4.52 and 4.40, respectively) suggests that restricted range could not explain the lack of significant findings for fathers' behavior. Finally, the analyses were conducted for males and females combined. It is possible that gender differences in the etiologies of problem behaviors exist which would necessitate tailoring of clinical interventions to additional characteristics of the target group (Eley et al., 1999; Langbehn et al., 1998).

Future Research

This study served as a limited exploration of potential mediating factors that might explain how the interaction between biological risk and adoptive parent psychopathology resulted in the expression of psychopathology by some adoptees but not others. Future research should focus on additional factors that might serve to moderate biological effects. For example, a recent study by Riggins-Caspers et al. (2000) showed that the strength of biological effects varied as a function of how much information about the physical and psychiatric status of the birth parent was disclosed to the adoptive parent. Examination of the association between biological risk, information disclosed to the adoptive parents, and parenting behaviors of the adoptive parents within these data showed that rates of both maternal and paternal caring behavior were lower among adoptive parents with an adoptee who had a biological risk for psychopathology and for whom no information about the biological parent was provided ($Z = -.5$ versus $Z = .1$ to $.2$). Maternal and paternal overprotective behaviors were elevated within this group as well ($Z = .5$ versus $Z = -.3$ to $.2$). Unfortunately, statistical analyses were not possible due to insufficient numbers.[4] If these effects are in fact reliable, then the examination of potential intermediary factors that account for this association should be examined. One likely focus might be parental attributions about the intent of adoptee misbehavior. Perhaps informing parents about the potential risk for psychopathology in their adoptee would make them less likely to engage in maladaptive attributions about the origins of the child's behavior and encourage parenting styles (e.g., warm and nonrestrictive) that would compensate for any biological risk for psychopathology (Alexander, Waldron, Barton, & Mas, 1989; Baden, & Howe, 1992; Bugental, & Shennum, 1984; Compas, Phares, Banez, & Howell, D. 1991; Smith, & O'Leary,

1995, 1998). The incorporation of additional behavioral and cognitive factors can be in the model tested–mediated moderation–and provide valuable information on the role of the child and the parent in the development of psychopathology. The identification of specific processes could then be used to intervene with families of high at risk for psychopathology.

Despite the above limitations, the paradigm used in this study demonstrates that the use of non-normative samples for identifying risk and protective factors relevant to the development of psychopathology continues to be a profitable approach (Luthar, Burack, Cicchetti, & Weisz, 1997; Rolf, Masten, Cicchetti, Nuechterlein et al., 1990). The adoption paradigm, while not a high-risk population in terms of traditional psychosocial risk, allows researchers to investigate the independent and synergistic effects of both biology and environment on behavior. Furthermore, the adoption paradigm allows researchers to examine the effects of multiple systems, both distal and proximal, on behavior by incorporating biology × environment and biology-environment correlation into a single theoretical model. Through this method, we will gain a better understanding of the mechanisms involved in the expression of behavior and from that develop preventive and intervention techniques.

NOTES

1. Because we included alcoholism as an indicator of genetic risk we have chosen to replace the term "genetic" with "biological" in order to acknowledge that there are potential intrauterine effects that have not been taken into account. Although intrauterine trauma is indeed an environmental effect, we include it as a biological effect because it alters the biological structure of the child.

2. Selective placement refers to the placement of a child into an adoptive home on the basis of similarities between some characteristic of the adoptive and birth parents (e.g., intelligence, personality).

3. Parent education and socioeconomic status were entered as covariates in the univariate analyses predicting parenting behaviors. No significant effects were found; consequently, these terms were removed from the model.

4. The analyses for the Riggins et al. (2000) paper combined two independent adoption studies to allow sufficient numbers.

REFERENCES

Alexander, J.F., Waldron, H.B., Barton, C., & Mas, C.H. (1989). The minimizing of blaming attributions and behaviors in delinquent families. *Journal of Consulting and Clinical Psychology*, 57, 19-24.

American Psychiatric Association (1987). *Diagnostic and Statistical Manual of Mental Disorder (3rd ed. revised)*. Washington, D.C.: American Psychiatric Association.

Anderson, K. E., Lytton, H., & Romney, D. (1986). Mothers' interactions with normal and conduct-disordered boys: Who affects whom? *Developmental Psychology, 22,* 604-609.

Baden, A.D., & Howe, G.W. (1992). Mothers' attributions and expectancies regarding their conduct-disordered children. *Journal of Abnormal Child Psychology, 20,* 467-485.

Baron, R.M. & Kenny, D.A. (1986). The moderator-mediator variable distinction in social psychological research: Conceptual, strategic, and statistical considerations. *Journal of Personality and Social Psychology, 51,* 1173-1182.

Bates, J., Pettit, G.S., Dodge, K.A., & Ridge, B. (1998). Interaction of temperamental resistance to control and restrictive parenting in the development of externalizing behavior. *Developmental Psychology, 34,* 982-995.

Bell, R. Q. (1968). A reinterpretation of the direction of effects in studies of socialization. *Psychological Review, 75,* 81-95.

Bell, R. Q. (1979). Parent, child, and reciprocal influences. *American Psychologist, 34,* 821-826.

Bell, R. Q., & Chapman, M. (1986). Child effects in studies using experimental or brief longitudinal approaches to socialization. *Developmental Psychology, 22,* 595-603.

Belsky, J., & Volling, B, L. (1987). Mothering, fathering, and marital interaction in the family triad during infancy: Exploring family system's processes. In P. W. Berman & F. A. Pedersen (Eds.), *Men's transitions to parenthood: Longitudinal studies of early family experience.* (pp. 37-63). Hillsdale, NJ: Lawrence Erlbaum.

Belsky, J., Hsieh, K., & Crnic, K. (1998). Mothering, fathering, and infant negativity as antecedents of boys' externalizing problems and inhibition at age 3 years: Differential susceptibility to rearing experience? *Development & Psychopathology, 10,* 301-319.

Bronfenbrenner, U. (1977). Toward an experimental ecology of human development. *American Psychologist, 32,* 513-531.

Bronfenbrenner, U. (1986). Ecology of the family as a context for human development: Research perspectives. *Developmental Psychology, 22,* 723-742.

Bronfenbrenner, U., & Ceci, S. (1994). Nature-nurture re-conceptualized in developmental perspective: A bio-ecological model. *Psychological Bulletin, 101,* 568-586.

Bugental, D. B., & Shennum, W. A. (1984). Difficult children as elicitors and targets of adult communication patterns: An attributional behavioral transactional analysis. *Monographs of the Society for Research in Child Development, 49* (1, Serial No. 205).

Cadoret, R.J., & Cain, C.A. (1981). Genetic-environmental interaction in adoption studies of antisocial behavior. In C. Perris, G. Struwe, & B. Jansson (eds.), *Biological Psychiatry* (pp. 97-100). Elsevier, North Holland: Biomedical Press.

Cadoret, R.J., Cain, C.A., & Crowe, R.R. (1983). Evidence for gene- environment interaction in the development of adolescent antisocial behavior. *Behavior Genetics, 13,* 301-310.

Cadoret, R., Yates, W., Troughton, E., Woodworth, G., & Stewart, M. (1995). Genetic-environmental interaction in the genesis of aggressivity and conduct disorders. *Archives of General Psychiatry, 52,* 916-924.

Cadoret, R., Winokur, G., Langbehn, D., Troughton, E., Yates, W., & Stewart, M. (1996). Depression spectrum disease I: The role of gene-environment interaction. *American Journal of Psychiatry, 153*, 892-899.

Campbell, S.B., Pierce, E.W., Moore, G., Marakovitz, S., & Newby, K. (1996). Boys' externalizing problems at elementary school age: Pathways from early behavior problems, maternal control, and family stress. *Development and Psychopathology, 8*, 701-719.

Caspi, A., Elder, G. H., & Bem, D. J. (1987). Moving against the world: Life-course patterns of explosive children. *Developmental Psychology, 23*, 308-313.

Compas, B. E., Phares, V., Banez, G. A., & Howell, D. C. (1991). Correlates of internalizing and externalizing behavior problems: Perceived competence, causal attributions, and parental symptoms. *Journal of Abnormal Child Psychology, 19*, 197-218.

Conger, R.D., Conger, K.J., Elder, G.H., Lorenz, F.O., Simons, R.L., & Whitbeck, L.B. (1992). A family process model of economic hardship and adjustment of early adolescent boys. *Child Development, 63*, 526-541.

Conger, R., Lorenz, F., Elder, G., Melby, J., Simons, R., & Conger, K. (1991). A process model of economic pressure and early adolescent alcohol use. *Journal of Early Adolescence, 11*, 430-449.

Coyne, J.C., & Downey, G. (1991). Social factors and psychopathology: Stress, social support, and coping processes. In M. R. Resenzweig and L. W. Porter (Eds.), *Annual review of psychology* (Vol. 42, pp. 401-425). Palo Alto, CA: Annual Reviews.

Cutrona, C.E., Cadoret, R., Suhr, J.A., Richards, C.C., Troughton, E., Schutte, K., & Woodworth, G. (1994). Interpersonal variables in the prediction of alcoholism among adoptees: Evidence for gene-environment interactions. *Comprehensive Psychiatry, 35*, 171-179.

Deater-Deckard, K. (2000). Parenting and child behavioral adjustment in early childhood: A quantitative genetic approach to studying family processes. *Child Development, 71*, 468-484.

Deater-Deckard, K., & Plomin, R. (1999). An adoption study of the etiology of teacher and parent reports of externalizing behavior problems in middle childhood. *Child Development, 70*, 144-154.

Dodge, K. A. (1990). Developmental psychopathology in children of depressed mothers. *Developmental Psychology, 26*, 3-6.

Downey, G., & Coyne, J.C. (1990). Children of depressed parents: An integrative review. *Psychological Bulletin, 108*, 50-76.

Elder, G.H., & Caspi, A. (1988). Economic stress in lives: Developmental perspectives. *Journal of Social Issues, 44*, 25-45.

Eley, T.C., Lichtenstein, P., & Stevenson, J. (1999). Gender differences in the etiology of aggressive and nonaggressive antisocial behavior: Results from two twin studies. *Adolescent Development, 70*, 155-168.

Feighner, J. P., Robins, E., Guze, S. B., Woodruff, R. A., JR., Winokur, G., & Munoz, R. (1972). Diagnostic criteria for use in psychiatric research. *Archives of General Psychiatry, 35*, 837-844.

Ge, X., Conger, R.D., Cadoret, R.J., Neiderhiser, J., Yates, W., Troughton, E., & Stewart, M. A. (1996). The developmental interface between nature and nurture: A mutual influence model of child antisocial behavior and parent behaviors. *Developmental Psychology, 32*, 574-589.

Greenwald, R. L., Bank, L., Reid, J. B., & Knutson, J. F. (1997). A discipline-mediated model of excessively punitive parenting. *Aggressive Behavior, 23*, 259-280.

Kendler, K.S. (1996). Parenting: A genetic-epidemiological perspective. *American Journal of Psychiatry, 153*, 11-20.

Kendler, K.S., Sham, P.C., & MacLean, C.J. (1997). The determinants of parenting: An epidemiological, multi-informant, retrospective study. *Psychological Medicine, 27*, 549-563.

Kochanska, G. (1997). Multiple pathways to conscience for children with different temperaments: From toddlerhood to age 5. *Developmental Psychology, 33*, 228-240.

Langbehn, D.R., Cadoret, R.J., Yates, W.R., Troughton, E.P., & Stewart, M.A. (1998). Distinct contributions of conduct and oppositional defiant symptoms to adult antisocial behavior. *Archives of General Psychiatry, 55*, 821-829.

Larzelere, R. E., & Patterson, G. R. (1990). Parental management: Mediator of the effect of socioeconomic status on early delinquency. *Criminology, 28*, 301-323.

Lempers, J., Clark-Lempers, D., & Simons, R.L. (1989). Economic hardship, parenting, and distress in adolescence. *Child Development, 60*, 25-39.

Lengua, L.J., Wolchik, S.A., Sandler, I.N., & West, S.G. (2000). The additive and interactive effects of parenting and temperament in predicting adjustment problems of children of divorce. *Journal of Clinical Child Psychology, 29*, 232-244.

Loney, J., Langhorne, J. E., Paternite, C. E., Whaley-Klahn, M. A., Blair-Broeker, C. T., & Hacker, M. (1980). The Iowa habit: Hyperkinetic/aggressive boys in treatment. In S. B. Sells, R. Crandall, M. Roff, J. S. Strauss, & W. Pollin (eds.), *Human Functioning in Longitudinal Perspective* (pp. 119-140). Baltimore, MD: Williams & Wilkins.

Luthar, S.S., Burack, J.A., Cicchetti, D., & Weisz, J.R. (1997). *Developmental psychopathology: Perspectives on adjustment, risk, and disorder.* Cambridge, UK: Cambridge University Press.

Lytton, H. (1982). Two-way influence processes between parents and child—when, where, how? *Canadian Journal of Behavioral Science, 14*, 259-275.

Lytton, H. (1990). Child effects—still unwelcome? Response to Dodge and Wahler. *Developmental Psychology, 26*, 705-709.

Maccoby, E. (1992). The role of parents in the socialization of children: An historical overview. *Developmental Psychology, 28*, 1006-1017.

Maccoby, E. E., & Martin, J. M. (1983). Socialization in the context of the family: Parent-child interaction. In G. H. Mussen (Series Ed.) & E. M. Hetherington (Volume Ed.), *Handbook of Child Psychology, Vol. 4: Socialization, personality, and social development.* New York: Wiley.

Maxwell, S.E., & Delaney, H.D. (1993). Bivariate median splits and spurious statistical significance. *Psychological Bulletin, 113*, 181-190.

O'Connor, T.G., Deater-Deckard, K., Fulker, D., Rutter, M., & Plomin, R. (1998). Genotype-environment correlations in late childhood and early adolescence: Antisocial behavioral problems and coercive parenting. *Developmental Psychology, 34*, 970-981.

Parker, G.H., Tupling, H., & Brown, L.B. (1979). A parental bonding instrument. *British Journal of Medical Psychology*, 52, 1-10.

Patterson, G. R. (1982). *A social learning approach, Vol. 3: Coercive family process.* Eugene, OR: Castalia Publishing Company.

Patterson, G.R. (1983). Stress: A change agent for family process. In N. Garmezy & M. Rutter (Eds.), *Stress, coping, and development in children* (pp. 235-264). New York: McGraw-Hill.

Patterson, G., DeBaryshe, B. D., & Ramsey, E. (1989). A developmental perspective on antisocial behavior. *American Psychologist*, 44, 329-335.

Plomin, R. & Bergeman, C. S. (1991). The nature of nurture: Genetic influence on "environmental" measures. *Behavioral and Brain Sciences*, 14, 373-427.

Plomin, R., DeFries, J.C., & Loehlin, J.C. (1977). Genotype-environment interaction and correlation in the analysis of human behavior. *Psychological Bulletin*, 84, 309-322.

Plomin, R., McClearn, G., Pederson, N., Nesselroade, J., & Bergeman, C. (1988). Genetic influence on childhood family environment perceived restrospectively from the last half of the life span. *Developmental Psychology*, 24, 738-745.

Reid, J.B., Patterson, G. R., & Loeber, R. (1982). The abused child: Victim, instigator, or innocent bystander. In D. Berstein (Ed.), *Response Structure and Organization* (pp. 47-68). Lincoln, NE: University of Nebraska Press.

Rey, J.M., & Plapp, J.M. (1990). Quality of perceived parenting in oppositional and conduct disordered adolescents. *Journal of the American Academy of Child and Adolescent Psychiatry*, 29, 382-385.

Riggins-Caspers, K., Cadoret, R., Knutson, J., & Langbehn, D. (2000). Gene-environment interaction and evocative gene-environment correlation: Contributions of harsh discipline and parental psychopathology to problem adolescent behaviors. Paper presented at the 30th Annual Behavior Genetics Association Meeting, June 29th-July 1st, Burlington, VT.

Robins, L., Helzer, J., Cottler, L., & Goldring, E. (1989). *NIMH Diagnostic Interview Schedule, Version 3, Revised (DIS-III-R).* St Louis, MO: Washington University School of Medicine.

Rolf, J., Masten, A., Cicchetti, D., Nuechterlein, K.H., & Weintraub, S. (1990). *Risk and protective factors in the development of psychopathology.* Cambridge, UK: Cambridge University Press.

Rothbaum, F., & Weisz, J.R. (1994). Parental caregiving and child externalizing behavior in nonclinical samples: A metanalysis. *Psychological Bulletin*, 116, 55-74.

Rowe, D. (1981). Environmental and genetic influences on dimensions of perceived parenting: A twin study. *Developmental Psychology*, 17, 203-208.

Rowe, D. (1983). A biometrical analysis of perceptions of family environment: A study of twin and singleton sibling kinships. *Child Development*, 54, 416-423.

Rutter, M. L. (1997a). Nature-nurture integration: The example of antisocial behavior. *American Psychologist*, 52, 390-398.

Rutter, M. (1997b). Integrating nature and nurture: Implications of person-environment correlations and interactions for developmental psychopathology. *Development and Psychopathology*, 9, 335-364.

Rutter, M., & Pickles, A. (1991). Person-environment interactions: Concepts, mechanisms, and implications for data analysis. In T. D. Wachs and R. Plomin, *Concepualization and measurement of organism-environment interaction* (pp. 105-141). Washington, DC: American Psychological Association.

Rutter, M., Dunn, J., Plomin, R., Simonoff, E., Pickles, A., Maughan, B., Ormel, J., Meyer, J., & Eaves, L. (1997). Integrating nature and nurture: Implications of person-environment correlations and interactions for developmental psychopathology. *Development and Psychopathology, 9*, 335-364.

Scarr, S., & McCartney, K. (1983). How people make their own environments: A theory of genotype → environment effects. *Child Development, 54*, 424-435.

Smith, A.M., & O'Leary, S.G. (1995). Attributions and arousal as predictors of maternal discipline. *Cognitive Therapy and Research, 19*, 459-471.

Smith, A.M., & O'Leary, S.G. (1998). The effects of maternal attributions on parenting: An experimental analysis. *Journal of Family Psychology, 12*, 234-243.

Stoolmiller, M. (1999). Implications of the restricted range of family environments for estimates of heritability and nonshared environment in behavior genetic adoption studies. *Psychological Bulletin, 125*, 392-409.

Stormshak, E.A., Bierman, K.L., McMahon, R.J., & Lengua, L.J. (2000). Parenting practices and child disruptive behavior problems in early elementary school. *Journal of Clinical Child Psychology, 29*, 17-29.

Susman, E., Trickett, P., Iannotti, R., Hollenbeck, B., & Zahn-Waxler, C. (1985). Child-rearing patterns in depressed, abusive, and normal mothers. *American Journal of Orthopsychiatry, 55*, 237-251.

Volling, B.L., & Belsky, J. (1992). The contribution of mother-child and father-child relationships to the quality of sibling interaction: A longitudinal study. *Child Development, 63*, 1209-1222.

Using the Stepfamily Genetic Design to Examine Gene-Environment Processes in Child and Family Functioning

Kirby Deater-Deckard
Judy Dunn
Thomas G. O'Connor
Lisa Davies
Jean Golding
and the ALSPAC Study Team

SUMMARY. Studying children in different types of families–intact, single-mother, and stepparent families–affords opportunities for testing

Kirby Deater-Deckard is affiliated with the University of Oregon. Professor Judy Dunn, Thomas G. O'Connor, and Lisa Davies are affiliated with the Institute of Psychiatry, London, UK. Professor Jean Golding is affiliated with Bristol University, UK.

Address correspondence to Kirby Deater-Deckard, Department of Psychology, 1227 University of Oregon, Eugene, OR 97403-1227 (E-mail: kirbydd@darkwing.uoregon.edu).

The authors thank the mothers who participated in the study and the midwives for their cooperation and help in recruitment.

This study was supported in part by the Medical Research Council, Wellcome Trust, Departments of Health and Environment, British Gas, and other companies.

The Avon Longitudinal Study of Parents and children (ALSPAC) is part of the European Longitudinal Study of Pregnancy and Childhood initiated by the World Health Organization. The ALSPAC Study Team comprises interviewers, technicians, clerical workers, research scientists, volunteers, and managers who continue to make the study possible. This work was conducted while the first author was at the Institute of Psychiatry, London, UK.

[Haworth co-indexing entry note]: "Using the Stepfamily Genetic Design to Examine Gene-Environment Processes in Child and Family Functioning." Deater-Deckard, Kirby et al. Co-published simultaneously in *Marriage & Family Review* (The Haworth Press, Inc.) Vol. 33, No. 2/3, 2001, pp. 131-156; and: *Gene-Environment Processes in Social Behaviors and Relationships* (ed: Kirby Deater-Deckard, and Stephen A. Petrill) The Haworth Press, Inc., 2001, pp. 131-156. Single or multiple copies of this article are available for a fee from The Haworth Document Delivery Service [1-800-HAWORTH, 9:00 a.m. - 5:00 p.m. (EST). E-mail address: getinfo@haworthpressinc.com].

models of gene-environment processes, based on estimates of sibling similarity among full-siblings and half-siblings. We used a stepfamily quantitative genetic design to estimate genetic and environmental sources of variance in children's behavior problems and prosocial behaviors, as well as negativity in their relationships with their mothers and mothers' partners. Participants included full- and half-sibling pairs (same- and opposite-sex) from the Avon Longitudinal Study of Parents and Children. Mothers reported on their children's behavior problems and prosocial behaviors, as well as negativity in their parent-child relationships, for a target child (4 years old) and one older sibling ($M = 6.31$ years). There was additive genetic variance in child behavior problems and partner-child negativity, and shared environmental variance in mother-child and partner-child negativity. One-fifth to two-thirds of the variance was accounted for by nonshared environment and error. These findings were similar even after controlling for sibling gender and age differences, the resident status of the older sibling, and the older siblings' degree of contact with the nonresident biological parent. The links between parental negativity and child behavior problems were mediated by genetic covariance suggesting possible gene-environment correlation processes, and the links between parental negativity and child prosocial behaviors were mediated primarily by environmental covariance. *[Article copies available for a fee from The Haworth Document Delivery Service: 1-800-HAWORTH. E-mail address: <getinfo@haworthpressinc.com> Website: <http://www.HaworthPress.com> © 2001 by The Haworth Press, Inc. All rights reserved.]*

KEYWORDS. Genetics, family environment, externalizing, internalizing, prosocial behavior

The link between positive and negative social behaviors on the part of children and the nature of those children's family environments has been well established–children who show problems in behavioral and emotional development are more likely to have relationships with parents that are antagonistic and negative. Behavioral genetic studies like those presented in this volume have shown that genetic and environmental factors combine in their influences on individual differences in social behaviors and the qualities of social relationships in families (Plomin, 1994a; Rutter et al., 1997). The majority of these studies have relied on samples of twins and adoptive family members. Our goal is to describe a complementary, yet underutilized, stepfamily genetic design using data from the Avon Longitudinal Study of Parents and Children (ALSPAC). In this genetic epidemiological study of family processes and children's behaviors, we compare sibling similarity in social-emotional adjustment and qualities of parent-child relationships across groups of full- and half-siblings residing in different types of families (i.e., intact, single-parent, step-parent).

GENETIC AND ENVIRONMENTAL INFLUENCES ON SOCIAL-EMOTIONAL ADJUSTMENT

Individual differences in child and adolescent social-emotional adjustment are thought to arise from a complex interplay between genetic and environmental factors. These individual differences include moderate to substantial environmental effects, with the majority being of the "nonshared" variety–that is, environmental influences that lead to sibling differences in adjustment and family relationships. By comparison, "shared" environmental influences–those that lead to sibling similarity–are more modest (Dunn & Plomin, 1990; Harris, 1998; Plomin, 1994b). Noteworthy exceptions to this general pattern of environmental influences on social-emotional adjustment include nonaggressive behavior problems and observations of aggressive behavior, which often include shared environmental effects (Eley, Lichtenstein, & Stevenson, 1999; Miles & Carey, 1997). Genetic factors also have been implicated, with additive genetic influences sometimes accounting for half or more of the variance, depending on the specific type and measure of social behavior or relationship process in question.

Findings from two prototypical behavior genetic studies, an adoption study and a twin study, serve to illustrate the method and findings in this literature. The Colorado Adoption Project (CAP) is an ongoing longitudinal study of approximately 100 pairs of unrelated adoptive siblings and 100 pairs of full biological siblings in matched comparison families (Plomin & DeFries, 1985). The adoptive sibling design provides a direct estimate of shared environmental effects–only shared environmental factors can be responsible for sibling similarity among biologically unrelated children reared in the same household. An indirect estimate of genetic variance can be obtained by comparing adoptive sibling similarity to biological sibling similarity in matched nonadoptive families. In the CAP, five annual assessments in middle childhood revealed that individual differences in parent- and teacher-rated behavior problems included genetic (17% to 49%) and shared environmental (0% to 27%) variance, although the majority of the variance (42% to 70%) was accounted for by nonshared environmental factors (Deater-Deckard & Plomin, 1999). In contrast, parent- and self-reported internalizing symptoms were not heritable, instead including modest shared environmental variance and substantial nonshared environmental variance (Eley, Deater-Deckard, Fombonne, Fulker, & Plomin, 1998).

A recent study of Swedish and British children and adolescents exemplifies the twin study approach (Eley et al., 1999). Using this design, the researchers derived indirect estimates of genetic and shared environmental variance by comparing the sibling similarity of identical twins (who share 100% of their genes) and fraternal twins (who share, on average, 50% of the genes that vary across individuals). Antisocial behavior problems (e.g., aggression and delinquency) were reported by parents for 1022 pairs of Swedish twins and 501 pairs of British twins. The majority of the variance in aggressive behavior problems was accounted for by genetic factors (50% to 70%); shared environment was modest (4% to 18%). In contrast, individual differences in nonaggressive behavior problems included both heritable (0% to 60%) and shared environmental (21% to 64%) sources of variance, with the added suggestion that the etiology may differ for boys and girls.

PROSOCIAL BEHAVIOR

Genetic research has focused primarily on indicators of maladjustment; by comparison, we know relatively little about gene-environment processes for positive, prosocial behaviors. This is a problematic gap in the literature, because positive attributes of children's behavior and their family environments are important aspects of social behavior and family functioning. Individual differences in prosocial behavior are negatively correlated with maladjustment, may serve to modulate the expression of behavior problems such as aggression and noncompliance, and along with supportive aspects of the family environment can buffer children from the deleterious effects of environmental risk factors such as poverty and harsh parenting (Eisenberg & Murphy, 1995; Pettit, Bates, & Dodge, 1997).

Two genetic studies illustrate this much smaller literature. Rushton and his colleagues (Rushton, Fulker, Neale, Nias, & Eysenck, 1986) administered a battery of questionnaires to a sample of adult twins (30 years old, on average), including an altruism checklist designed for their study, an emotional empathy scale (Mehrabian & Epstein, 1972), and a nurturance scale from the Personality Research Form (Jackson, 1974). Heritability was moderate to substantial (.56 to .70) for all three self-reported indicators of prosocial behavior in adulthood (i.e., altruism, empathy, and nurturance), and shared environment was negligible.

In a longitudinal study of twins at 14 months and 20 months (Zahn-Waxler, Robinson, & Emde, 1992), each twin was observed following an episode where his or her mother feigned a sore knee. Prosocial actions, empathic concern for the mother, and indifference were moderately heritable (.25 to .35), and shared environment was negligible. In contrast, mothers' reports of their twins' prosocial behaviors were moderately heritable (.34 at 14 months and .36 at 20 months), but also included shared environmental variance (.49 at 14 months, .45 at 20 months). There are relatively few behavioral genetic studies examining prosocial behaviors, and the methods differ across studies. Much more research is needed that addresses gene-environment processes for prosocial as well as antisocial behaviors in childhood.

GENE-ENVIRONMENT PROCESSES
IN FAMILY RELATIONSHIPS

Although informative, studies that estimate genetic and environmental influences on a developmental outcome such as behavior problems or prosocial behavior do not reveal how these influences actually work together, nor do they identify those genetic and environmental variables that actually predict individual differences. For instance, discovering that shared or nonshared environmental influences exist for prosocial behavior does not help us identify which or how many environmental factors are operating. In addition, the methods and findings described above do not identify how genetic and environmental factors may operate together. In order to do this, genetic studies that include measures of the children's environments as well as the children's behaviors are required.

To this end, we also examined genetic and environmental sources of variance in a measure of parent-child negativity and conflict, because previous research has demonstrated the importance of this particular factor in the development and maintenance of problems in social-emotional development (Rohner, 1986; Patterson, 1997). Parent-child negativity may operate as an environmental influence and may also operate as part of a gene-environment process. For instance, parental negativity toward a child may promote behavior problems because the child interprets this parental behavior as rejection (Rohner, 1986). This process may operate differentially *between* families, with siblings residing together being more similar in their social-emotional adjustment compared to children in other families who are exposed to different levels of parental negativity. This would be an example of a shared environmental process. At the same time, parental negativity may operate differentially *within* families, whereby higher levels of parental negativity toward one child are associated with that child having more behavioral and emotional problems and fewer prosocial skills than his or her sibling who is treated more favorably (Kowal & Kramer, 1997). This would be an example of a nonshared environmental process. It is noteworthy that a single environmental factor like parental negativity can have both shared and nonshared environmental influences.

The above discussion presumes that parental negativity operates only as an environmental influence, independent of child genetic factors. However, it is likely that this as well as other environmental factors operate through transactions with (rather than independent of) genetic influences. Evocative gene-environment correlation is one such process, whereby an individual's genetically influenced behaviors or attributes predispose him or her to certain experiences, which in turn serve to reinforce and maintain that behavior or attribute (Scarr & McCartney, 1983). For example, a child who is highly prosocial and shows few antisocial behaviors or emotional problems is likely to elicit warmer and more positive behaviors from her or his parents, compared to another child who is not prosocial, is more difficult to manage behaviorally and is more emotionally labile. There may be genetic influences operating on these prosocial and antisocial behaviors, but in this case the genetic influences operate in part or entirely through their impact on the child's experiences in relationships with her or his parents. The first step toward identifying gene-environment correlation is to examine whether children's genetic factors influence the environmental factor of interest—in this study, parental negativity.

THE STEPFAMILY GENETIC DESIGN

We examined full-sibling and half-sibling similarity using data from a large and diverse epidemiological community study of "intact" or nonstep, step-, and single-parent families. The similarity of resident full-siblings (who share half of their individual differences genes, on average) is compared to the similarity of resident half-siblings (who share 25%, on average). Genetic influences are implicated if full-sibling similarity exceeds half-sibling similarity. At the same time, environmental influences are implicated when full- and half-sibling similarity is comparable, and when sibling differences are not accounted for by genetic variation. See Petrill (this volume) for details regarding the quantitative genetic design.

The epidemiological stepfamily method has some advantages that make it an ideal complement to twin and adoption study designs. First, siblings who share the same genetic mother are more likely to have had similar prenatal environments than are two unrelated siblings. Thus, sibling similarity in the prenatal environment for twins, full siblings, and unrelated adoptive siblings is probably confounded with genetic similarity and may bias estimates of genetic and environmental variance. When identical and fraternal twins are compared, the sibling similarity of the prenatal environment may be confounded with the genetic similarity of the twins, because identical twins are more likely to share a chorion than are fraternal twins (Plomin, DeFries, McClearn, & Rutter, 1997, p. 81). When unrelated adoptive siblings (two adopted siblings who had different genetic mothers) and biological full-siblings are compared, sibling similarity in the prenatal environment and preadoptive environment may be confounded with genetic similarity. Full siblings, though born at different times, probably have more similar prenatal environments than do unrelated adopted siblings who had different genetic mothers. Although the degree to which this confound influences estimates of genetic and environmental variance in twin and adoption studies is debated (Devlin, Daniels, & Roeder, 1997; Sokol et al., 1995), one advantage of the stepfamily genetic design is that this confound is minimized; in most cases, resident full-siblings and half-siblings share the same genetic mother and therefore have similarly correlated prenatal environments.

A second advantage to the stepfamily genetic design is one of sample representativeness and the generalizability of findings. "Intact" or nonstep families with full-siblings and step- and single-parent families with full-siblings and half-siblings are common, together representing the vast majority of families (Haskey, 1994; Amato, 1999). In Britain (the location of the current study), the modal family structure includes two full-siblings who reside with both of their biological parents. However, full-sibling and half-sibling pairs living with their mothers and stepfathers are also common. In the UK, 1 in 8 children will live in a stepfamily before she or he is 16 years old, and most of these children will spend some time living in a single-parent (usually mother-headed) household during their parents' transitions between partnerships (Haskey, 1994). By comparison, the twin birth rate is approximately 1 in 85 births (Plomin et al., 1997, p. 81), and adoptive families are relatively rare. Importantly, the findings from twin and adoption studies should not be disregarded on this basis. On the contrary, the designs are complementary, and the findings from all genetically informative family studies should be considered together when testing theories regarding genetic and environmental processes in development.

Third, nonstep, step- and single-parent families are found across the full range of socioeconomic circumstances. In contrast, disadvantaged and highly stressful family environments are typically underrepresented in adoptive samples because adoptive families are selected on the basis that they will provide sufficient resources and a loving family environment. This "restricted range" in family environments in adoptive families has been raised as a potential source of bias in genetic and environmental parameter estimates derived from adoption studies (Stoolmiller, 1998). Environmental range restriction is less of a concern in community studies of full- and half-siblings.

Fourth and finally, the main difficulty with the stepfamily genetic design also turns out to be a potential strength. The disadvantage to examining such a small genetic difference between half- and full-siblings is that larger samples are required in order to have sufficient statistical power. At the same time, by examining adjustment among full- and half-sibling pairs that have experienced changes in family structure and functioning (e.g., divorce and remarriage, losing and regaining contact with nonresident fathers, relocating, changes in custody arrangements, unpredictable fluctuations in economic resources), the impact of these family contextual factors on estimates of genetic and environmental variance estimates can be ascertained. Thus, as long as statistical power is sufficient, one gains much by being able to examine not only gene-environment processes, but also family contextual influences on the estimation of those genetic and environmental parameters.

We must emphasize the importance of incorporating stepfamily genetic designs into the battery of tools used for research on genetic and environmental influences. As noted above, there is a need to replicate findings from twin and adoption studies using family designs that include full-siblings and half-siblings living in intact as well as single-parent and stepfamilies. All of these behavior genetic models assume that the quantitative genetic approach will work equally well regardless of the specific type of design that one is using—for instance, comparing twins, comparing adoptive siblings, or comparing full- and half-siblings. Nearly all of the research in this literature has relied on twin and adoption designs. The stepfamily design provides a much-needed test of the generalizability of the findings derived from twin and adoption studies.

Though used far less frequently than twin and adoption studies, there are several examples of studies using this stepfamily genetic design. The Nonshared Environment in Adolescent Development (NEAD) project was a longitudinal study of approximately 400 sibling pairs–some were twins, while others were full-siblings, half-siblings, or unrelated siblings, all living in intact or reconstituted stepparent families (Reiss et al., 1994). In NEAD, the stepfamily design was embedded within a combined twin and adoptive sibling design, and all of the stepfamilies were headed by married couples who had been together for 9 years, on average. The investigators were able to examine the role of sibling genetic similarity in predicting sibling similarity in adjustment by representing the entire range of sibling genetic similarity in the population– from genetically identical twins, to fraternal twins and full-siblings in intact and stepfamilies (who share, on average, 50% of their genes), to half-siblings in stepfamilies (who share, on average, 25% of their genes), to unrelated siblings in stepfamilies.

Another approach is represented by the analysis of data on extended kinships in large community studies, such as the work by Rowe and his colleagues on data from the National Longitudinal Study of Youth or NLSY (Rodgers, Rowe, & Li, 1994; Van den Oord & Rowe, 1997). In these studies, data for siblings of varying genetic similarity (sometimes twins, sometimes full-siblings, sometimes half-siblings) as well as cousins were incorporated into analyses examining genetic and environmental sources of variation in maladjustment. In one analysis (Rodgers et al., 1994), the genetic and environmental sources of variance in parent-rated child behavior problems were similar in combined cousin/sibling genetic models regardless of whether twin pairs were included or excluded.

These studies provide a model upon which we have based the current study. We have focused on data for full- and half-siblings derived from a large epidemiological community study. We included both single- and two-parent households, and stable (i.e., no partnership changes since birth of younger child) and unstable (i.e., one or more partnership changes) stepfamilies, with the goal of maximizing the diversity and representativeness of the sample.

METHOD

Sample

The sample included 3,503 pairs of siblings–a four-year-old target child, and the next older sibling–living in families who were participating in the Avon Longitudinal Study of Parents and Children (ALSPAC), an ongoing epidemiological study of nearly 14,000 women who were pregnant and gave birth to the "target" child during a 21-month period (from April 1 1991 to December 31 1992) in Avon county, England (Golding, Pembrey, Jones & ALSPAC Study Team, 2001). About 85% of the eligible population participated. In terms of sociodemographic indicators such as parental education and household structure, this large community sample is representative of Avon county (which includes the city of Bristol), and the sample also resembles the British population generally with the exception of a lower number of ethnic minority groups (4%). Attrition has been low–the majority (75%) of the study participants have remained in the study through its first five years.

For the quantitative genetic analyses, siblings were defined according to their biological relatedness. Full-siblings had the same biological mother and father, whereas half-siblings had the same biological mother but different biological fathers. Families were also defined according to the household composition and mothers' partnership status. Intact or nonstep families were two-parent families where the parents were the biological parents of all the children living in the household. Stepfamilies were two-parent families where one or more of the children were stepchildren of the resident partner or mother. Single-mother families were those families where the mother reported not having a resident partner, although some of the single mothers did have nonresident partners. Thus, intact nonstep families included only full-siblings (n = 2904 pairs), whereas step- and single-mother families included full-siblings (n = 359 pairs) and half-siblings (n = 240 pairs). The aim of this study was to examine within-family variability (i.e., sibling differences). The analysis of between-family variability (i.e., family-type differences) in the children's social-emotional adjustment is presented elsewhere (Dunn, Deater-Deckard, Pickering, O'Connor, 1998).

A description of the sibling pairs is provided in Table 1. Older siblings were 5.80 years old (on average) among full-siblings in nonstep families, 6.53 years old among full-siblings in step- and single-mother families, and 9.47 years old among half-siblings in step- and single-mother families. For all three groups of siblings, about one-quarter

TABLE 1. Description of sample.

| | Step- and Single-Mother Families | | Intact Families |
	full-siblings (359 pairs) % or M (SD)	half-siblings (240 pairs) % or M (SD)	full-siblings (2904 pairs) % or M (SD)
older sibling age in years	6.53 (3.8)	9.47 (3.9)	5.80 (2.9)
% pairs same-sex males	26%	27%	25%
% pairs same-sex females	23%	23%	25%
% pairs opposite-sex	51%	50%	50%
% living with married parents	39%	50%	95%
% living with cohabiting parents	15%	25%	5%
% living in step-father families (n = 340)	53%	47%	na
% living in other step families (n = 43)	47%	54%	na
% living in single-mother families (n = 216)	74%	26%	na
% partner not biological father of target child	45%	67%	100%
age of target child when father left (if applicable)	22.26 (15.48)	12.40 (15.81)	na
% older sibling lives in household	88%	94%	100%
days/month older sibling lives in household if joint custody	11.15 (10.66)	13.08 (10.34)	na
% older siblings with no contact with other parent	5%	36%	na

Note: na = not applicable.

were same-sex males, one-quarter same-sex females, and half opposite-sex pairs. Nearly all of the parents of full-siblings in intact nonstep
families were married as opposed to cohabiting, compared to half for
half-sibling pairs and about one-third for full-sibling pairs in stepfamilies.
Nearly one-half of the full-siblings who did not reside in intact nonstep
families, and one-quarter of all half-siblings, lived with a single mother.
Although the proportions of full- and half-sibling pairs were approximately equal among stepfamilies, there were a disproportionate number
of full-siblings in single-mother families (74%) compared to half-siblings in single-mother families (26%).

About one-half to two-thirds of the stepfamilies included a partner
who was not the biological parent of the 4-year-old target child–that is,
the mother had changed partners since the pregnancy. Within these and
the single-mother families, target children in full-sibling pairs were
about 22 months old when the biological father left the household,
whereas the target children in half-sibling pairs were about 12 months
old when the biological father left. Nearly all of the older siblings resided permanently in the same household as the target child. Of those

older siblings who lived in two households, they lived for one-third of the month in the target child's household, on average. Among the step- and single-mother families, few older full-siblings (5%) had no contact with their nonresident biological parent, whereas this proportion was larger (36%) for older half-siblings.

Procedure

Mothers completed a mailed questionnaire when the target child was 4 years old. This questionnaire included questions regarding the target child's behavior and emotional adjustment, the mother's and her partner's feelings toward the target child, and household structure. Mothers also completed questions about the behavioral and emotional adjustment of the next older sibling, if applicable.

Measures

Mothers rated both children's behavior and emotional problems and prosocial behaviors over the prior six months on the Strengths and Difficulties Questionnaire or SDQ (Goodman, 1997), a newly developed instrument that has been validated against the Child Behavior Checklist (Goodman & Scott, 1999). Mothers rated the truthfulness of 25 statements using a three-point Likert-type scale (0 = not true, 1 = somewhat true, 2 = certainly true). We used the twenty-item Total Problems score (aggression, conduct problems, hyperactivity, peer relationship problems, and emotional problems, $\propto$ = .80), and the 5-item Prosocial Behavior score $\propto$ = .80) that includes indicators of positive social behaviors such as sharing and kindness. Each mother also rated her own negativity toward each child, as well as that of her current partner, using a 4-item scale developed for this study (see Dunn et al., 1998). Higher scores on these scales indicated greater hostility toward that particular child.

In this study, we have focused on estimates of sibling similarity; the family and sibling type mean differences in these child and parenting behaviors are described elsewhere (Dunn et al., 1998). It is worth noting that the SDQ distribution characteristics for this sample were representative of the range of normal individual differences found for preschool and school-age children in the United Kingdom (see Goodman, 1997; Goodman & Scott, 1999).

RESULTS

Sibling intra-class correlations for mother-rated child SDQ Total Problems and SDQ Prosocial Behavior are presented in Table 2 as a function of sibling type (full-siblings in intact nonstep families, full-siblings in step- and single-mother families, and half-siblings in step- and single-mother families). In general, full-siblings were rated as being more similar in SDQ Total Problems than were half-siblings. In contrast, full- and half-siblings were similarly correlated for SDQ Prosocial Behaviors. The intra-class correlations for mother-rated mother-child negativity and partner-child negativity were also estimated and are shown in Table 2. Sibling similarity for mother-child negativity was comparable for all three types of siblings, whereas for partner-child negativity, full-sibling similarity was somewhat greater than half-sibling negativity.

Before proceeding with analyses, we examined whether gender and the age difference between the siblings had systematic influences on sibling similarity in mothers' ratings of child social behaviors and negativity in the parent-child relationship. There was no evidence to suggest gender and age differences in the estimates of sibling similarity, so we used the original uncorrected scores for all subsequent analyses.[1]

We were also able to examine two other factors that might systematically influence half-sibling similarity, including the resident status of the older sibling and contact between the older sibling and her or his nonresident parent (usually the father). It is possible that half-siblings

TABLE 2. Sibling intra-class correlations with sample sizes in parentheses, as a function of sibling type (full-siblings or half-siblings) in different family types (intact or other = single-mother, step-parent).

	Intact, Full sibs	Other, Full sibs	Half sibs
Child Behavior			
SDQ: Total problems	.38 (2836)	.31 (347)	.18 (236)
SDQ: Prosocial behavior	.25 (2870)	.25 (355)	.23 (237)
Parent-child relationship			
Maternal negativity toward child	.45 (2671)	.36 (325)	.43 (218)
Partner negativity toward child	.53 (2696)	.53 (227)	.42 (176)

Note: all correlations significant (two-tailed p < .05).

would be rated more similarly if they resided together, compared to half-siblings who did not reside together. It also is possible that half-sibling similarity would be greater among those pairs where the older half-sibling spent no time with her or his nonresident biological parent, compared to those who did spend time with the nonresident biological parent.

We considered the effects of these residency and visitation variables on estimates of half-sibling similarity using partial correlations. These were compared to the zero-order sibling intra-class correlations, as reported in Table 2. First, we estimated intra-class partial correlations for half-siblings again, this time excluding 14 families where the older sibling did not reside with the younger sibling. Dropping these families from the analyses had no impact on estimates of half-sibling similarity: SDQ Total Problems $r = .17$, $n = 222$ pairs; SDQ Prosocial, $r = .22$, $n = 223$; maternal negativity $r = .43$, $n = 206$; partner negativity $r = .41$, $n = 166$. Second, we estimated sibling intra-class partial correlations for half-siblings again, this time including only those families where the older sibling had no contact with the nonresident biological parent ($n = 87$ families). Including only those families where the older half-sibling had no contact with her or his nonresident biological parent had little impact on estimates of sibling similarity: SDQ Total Problems $r = .23$, $n = 86$ pairs; SDQ Prosocial, $r = .35$, $n = 86$; maternal negativity $r = .40$, $n = 79$; paternal negativity $r = .37$, $n = 68$. Therefore, we did not statistically control for these factors in our analyses.

Next, genetic and environmental effects were estimated using the ACE model (see Figure 1). In this standard model, measured variables are represented as rectangles, latent or unmeasured variables are represented as circles, regression paths are represented as one-headed arrows, and correlations are represented as two-headed arrows. The correlation between each sibling child's score on a measure (i.e., child 1 and child 2) is decomposed into three independent latent factors: additive genetic effects (A), shared environment (C), and nonshared environment + error (E) (Neale & Cardon, 1992). Additive genetic effects represent variation that is attributable to genetic influences, shared environment represents environmental influences that lead to sibling similarity, and nonshared environment represents environmental influences that lead to sibling differences. The effect of each latent factor is estimated twice—once for each child within the sibling pair (e.g., A1 and A2 for child 1 and child 2 respectively). The path or regression coefficients in the model (LA, LC, and LE) are used to estimate the proportions of variance (i.e., R^2) that are attributable to genetic and environmental ef-

FIGURE 1. The univariate ACE structural equation model for estimating additive genetic (A), shared environmental (C), and nonshared environmental + error (E) sources of variance. See text for a description of this model.

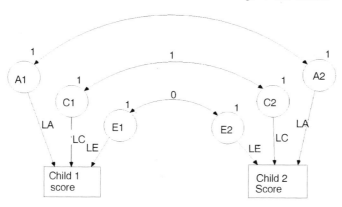

0.5 for full-siblings in intact families
0.5 for full-siblings in other families
.25 for half-siblings in other families

fects. There are three types of variance estimates: genetic variance or heritability (h^2), shared environmental variance (c^2), and nonshared environmental variance (e^2). Together, all of the variance in the measure is accounted for by these three effects, so that $h^2 + c^2 + e^2 = 1$.

In this study, we examined three types of sibling pairs: full-siblings in intact families, full-siblings in single parent and stepfamilies, and half-siblings in single parent and stepfamilies. Accordingly, the correlation between the two additive genetic variables (A1 and A2) was fixed at .5 for full-siblings and .25 for half-siblings, because full-siblings share 50% of their individual differences genes, on average, and half-siblings share 25% of their individual differences genes, on average. The correlation between the two shared environment latent variables (C1 and C2) was 1 for full- and half-siblings, because shared environment influences are, by definition, common to all siblings regardless of their genetic similarity. Finally, the correlation between the two nonshared environment latent variables (E1 and E2) was fixed at 0 for all siblings, because nonshared environmental factors (including error) are unique to each child. The ACE model includes several assumptions that are discussed in detail elsewhere (Loehlin, 1992)–there is no assortative mating, shared and nonshared environmental effects are equal across sibling types, genetic effects are additive, and gene-environment interaction and correlation is negligible.

The results are presented in Table 3. First, we estimated the genetic and environmental sources of variance in children's SDQ Total Problems scores. This model included additive genetic and nonshared environmental influences. Heritability accounted for the majority of the variance, and shared environment was negligible. Second, we estimated these parameters for children's SDQ Prosocial Behavior. There was a modest shared environment effect, but most of the variance (about two-thirds) was due to nonshared environment and error. Third, we examined sources of variance in maternal negativity toward each child. In this model, shared and nonshared environment were prominent, accounting for one-third and one-half of the variance respectively. Fourth, we estimated parameters for her partner's negativity toward each child. This model included statistically significant genetic and shared environment parameters, accounting for one-half and one-quarter of the variance respectively.

In order to test whether the family context (i.e., intact two-parent households compared to single parent and step families) influenced these estimates of genetic and environmental influence, we dropped full-siblings in intact families from the analyses so that only full- and half-siblings in single parent or step families were included. The variance estimates for this selected sample were: SDQ Total Problems, h^2 =

TABLE 3. Additive genetic (LA), shared environment (LC), and nonshared environment + error (LE) parameter estimates, univariate variance estimates for heritability (h^2), shared environment (c^2) and nonshared environment (e^2), and model fit indices including χ^2, p-value, and goodness of fit index (GFI).

	Parameter estimates			Variance estimates (%)			Model fit indices		
	LA	LC	LE	h^2	c^2	e^2	χ^2	p	GFI
Total problems	.87	.00	.49	76***	00	24	21.48	.01	.96
Prosocial behavior	.39	.42	.82	15	18*	67	1.69	.95	.99
Mother negativity	.43	.59	.68	18	35***	47	5.69	.46	.99
Partner negativity	.74	.50	.45	55***	25***	20	1.96	.92	.99

Note: * p < .05, *** p < .001 (p-values reported for variance estimates only). Because any estimate of nonshared environmental variance (e^2) includes error variance which is known to be > 0, it would not be appropriate to test the hypothesis that e^2 > 0. Therefore, the statistical significance for the nonshared environment parameter (LE) was not estimated.

.61, c^2 = .01, e^2 = .38; SDQ Prosocial behavior, h^2 = .13, c^2 = .18, e^2 = .69; maternal negativity, h^2 = .00, c^2 = .40, e^2 = .60; partner negativity, h^2 = .61, c^2 = .24, e^2 = .15. Model fit was very good; p-values were all greater than .35, and goodness of fit indices were in .99 to 1.0 range. However, given the smaller sample size in the analyses of the selected sample, few of these parameters were statistically significant. Nonetheless, effect sizes (i.e., parameter estimates of genetic and environmental sources of variance) based on this selected sample were very similar to the parameter estimates based on the full sample, as reported in Table 3.

In the final set of analyses, we wanted to examine the links between parental negativity and children's behavioral adjustment, in order to test for potential gene-environment and nonshared environmental processes that might mediate these associations between parent and child behavior. First, phenotypic bivariate correlations were estimated. Higher maternal negativity was associated with higher SDQ Total Problems, r = .50, n = 6540, p < .001, and lower Prosocial scores, r = −.28, n = 6579, p < .001. Partner negativity showed a similar pattern for Total Problems, r = .46, n = 6312, p < .001, and for Prosocial scores, r = −.25, n = 6350, p < .001. These correlations were similar even when computed separately for the 4-year-old target children and the older siblings (r = .45 to .51 between mother or partner negativity and child SDQ Total Problems, r = −.22 to −.34 for mother of partner negativity and child SDQ Prosocial scores).

A bivariate extension of the univariate ACE model was used to estimate the proportions of genetic and environmental mediation of the associations between parent and child behavior (see Pike, McGuire, Hetherington, Reiss et al., 1996). Because the findings for the univariate estimates were generally similar across the alternate models, a simplified three-group model (FI, FS, HS) based on all families was used to derive these estimates. In the bivariate model, the covariances between child behavior (SDQ Total Problems or Prosocial scores) and parent behavior (maternal or partner negativity) are decomposed into mediating (i.e., common or overlapping) and unique (i.e., specific to child behavior or parent behavior) genetic and environmental components through the estimation of six latent variables: A, C, and E that mediate parent and child variables, and A, C, and E that are unique to the parent and child variables.

The correlation between SDQ Total Problems and parental negativity was accounted for entirely by genetic mediation (96% for maternal negativity, and 100% for partner negativity); for maternal negativity, χ^2 = 46.73, 23 df, p < .003, goodness of fit index (GFI) = .95, and for partner negativity, $\chi^2 = 28.06$, 23 *df*, *p* = .21, GFI = .97. The correlation with maternal negativity also included negligible nonshared environmental mediation of 4%. In both of these analyses, shared environmental variance in SDQ Total Problems scores was constrained to equal 0 because the univariate analyses had already demonstrated that there was no shared environmental variance in SDQ Total Problems.

In contrast, the pattern of genetic and environmental mediation was different for children's Prosocial behaviors. The correlation between SDQ Prosocial scores and maternal negativity was mediated nearly equally by shared (58%) and nonshared (42%) sources of covariance, $\chi^2 = 9.94$, 21*df*, *p* = .98, GFI = .99. The correlation between SDQ Prosocial scores and partner negativity was mediated by nearly equal parts of genetic (33%), shared environmental (29%), and nonshared environmental (38%) covariance, although in this final model none of these mediating effects was statistically significant, $\chi^2 = 4.41$, 21*df*, *p* = 1, GFI = 1. Thus, child genetic sources of covariance entirely explained the correlation between children's behavior problems and parental negativity, whereas the sources of covariance between children's prosocial behaviors and parental negativity were mediated primarily by environmental sources of covariance.

DISCUSSION

Quantitative genetic models like the stepfamily design are useful for identifying some of the gene-environment processes that may be operating in the link between parenting environments and children's social-emotional development. In this study, we used the stepfamily genetic design to estimate genetic and environmental influences on children's behavior problems, prosocial behaviors, and their parents' negativity toward them, by comparing 3,263 full-siblings in intact, stepparent and single mother families and 240 half-siblings in stepparent and single mother families. The study revealed that mothers' reports of their children's positive and negative behaviors, as well as the quality of their relationships with their parents, were familial; siblings were modestly to moderately similar (intraclass *r* = .18 to .53). Sibling similarity was not affected when between-family differences in the children's ages, sexes, and residency or visitation patterns were statistically controlled.

Adjustment Problems and Prosocial Behaviors

Distinct patterns of genetic and environmental effects were found for adjustment problems and prosocial behavior. Individual differences in behavior and emotional problems were substantially heritable–additive genetic influences accounted for two-thirds to three-quarters of the variance. Shared environmental variance was negligible and nonsignificant, suggesting that environmental influences did not predict sibling similarity in adjustment problems. However, nonshared environmental variance was moderate, implicating environmental influences that lead to sibling differences in adjustment problems. This pattern of findings–that of substantial genetic and modest shared environmental influences–is consistent with other genetic studies using questionnaire-based parent-report measures (Deater-Deckard & Plomin, 1999; Eley et al., 1999; Rodgers et al., 1994).

In contrast to the findings for behavior and emotional problems, genetic as well as shared environmental influences were modest for children's prosocial behaviors, with the majority of the variance being accounted for by nonshared environmental variance (which includes measurement error). The lack of heritable and shared environmental variance in prosocial behavior contradicts some previous genetic research (Rushton et al., 1986; Zahn-Waxler et al., 1992). At the same time, all of these studies found moderate to substantial nonshared environmental variance in measures of prosocial behavior, suggesting that the findings in the current study are probably not simply a result of measurement error.

That environmental influences (be they nonshared or shared) may be operating on prosocial behavior is not surprising, given that individual differences in children's prosocial behaviors have been shown to be related to various attributes of the family environment such as parenting style (Krevans & Gibbs, 1996) and feeling-state talk (Dunn, Brown, & Beardsall, 1991). Differential treatment of children within the same family is implicated by the substantial nonshared environment effect, suggesting that these environmental influences may make siblings different from one another in prosocial behaviors. More research on prosocial behaviors is needed, as there are only a handful of genetic studies with each using different methods (e.g., self-report, parents' ratings, observations) and focusing on different age groups (e.g., toddlers, school-age children, adults).

Parental Negativity

We also examined the mother's reports of her negativity toward each child, and her perceptions of her current partner's negativity toward each child. Our purpose for examining this aspect of the family environment using a genetic design was to test whether parental negativity may be operating as a shared environmental factor, a nonshared environmental factor, or as part of a gene-environment correlation process. There was clear evidence that mothers' reports of parental negativity included shared and nonshared environmental variance. Over half of the variance in maternal negativity was environmental–about one-half of this was shared, and the other half was nonshared. A similar pattern of environmental variation was found for partner negativity. Other studies have also found evidence for shared and nonshared environmental variance in parent-report measures of parental negativity (Deater-Deckard, 2000).

Parental negativity also may be part of an evocative gene-environmental correlation process, whereby the child's behavior elicits more negativity from the parent. If this is true, parental negativity would be child-specific within families–that is, the mother's negativity toward one child would differ from her negativity toward another child. Furthermore, this differential treatment of sibling children must be greater among siblings who are genetically less similar. There was some evidence for this, as significant genetic variance was found in mothers' reports of their partners' negativity toward the children, and more importantly, much of the phenotypic correlation between parental negativity and child behavior problems was mediated by child genetic sources of covariance. This finding is consistent with other genetic studies of parent-child negativity (Deater-Deckard, 2000; Ge et al., 1996; O'Connor, Hetherington, Reiss, & Plomin, 1995; O'Connor, Deater-Deckard, Fulker, Rutter, & Plomin, 1998). Interestingly, this pattern of results was not found for children's prosocial behaviors, suggesting that these evocative child effects may be most salient (or, possibly only present) for noxious child behaviors such as aggression, negative mood, and noncompliance.

Another important contrasting pattern of results emerged in our comparison of maternal and mother-reported partner negativity. Child genetic variance was modest and nonsignificant for mothers' reports of their own negativity, but was moderate for mothers' reports of their partners' negativity. One explanation for this finding is that parents may be biased toward reporting similar treatment of their sibling children with respect to their own behavior, but this bias may be less pronounced (or not present) when the parents are reporting the behavior of their partners. If this bias is present, and there are child genetic influences on parenting behavior, then sibling similarity in the parental negativity they receive will be independent of sibling genetic similarity only for parents' self-reports (in contrast to parents' reports of the other parent). This is what we found in the current study. These findings point toward the importance of utilizing multiple methods and respondents in order to more readily discern the influence of such rater biases.

There are several limitations that should be considered when interpreting these findings. The most obvious concern is that we have relied on mothers' reports alone. These findings may not generalize to teachers' ratings, children's self-reports, or observational assessments of maladjustment and prosocial behavior. For instance, there is evidence from twin and adoption studies that heritable variance in behavior problems is lower in observational assessments compared to questionnaire-based assessments (Miles & Carey, 1997). Another limitation is that by relying on general indicators of family instability (e.g., changes in partnerships, current family structure), we were not able to ascertain more precisely the impact of specific family processes (e.g., parental conflict, family cohesiveness) on sibling similarity. In addition, nearly all of the participants in this study were Caucasian; the findings from this British study may not generalize to other ethnic and national populations.

CONCLUSIONS

The stepfamily genetic design is a powerful method for examining genetic and environmental influences on social behavior, and like the twin and adoption designs, can also be used to test for gene-environment correlation involving various aspects of the parent-child relationship and parenting environment. Many family and marital researchers could address questions regarding genetic and environmental influences on a wide variety of individual and relationship factors—indeed, some have existing data that could be examined using the method we have described here. Nonetheless, the design is not without its potential shortcomings. Most notably, changes in family structure and functioning that arise from parental separation and divorce, single parenthood, and remarriage, could affect sibling similarity through changes in residency and contact with parents and fluctuations in differential parental treatment. We examined this possibility in these data, and this does not appear to be the case in this study. The pattern of findings–additive genetic effects accounting for all of the sibling similarity in behavior problems, and additive shared environmental effects accounting for nearly all of the sibling similarity in prosocial behavior–was consistent across two tested variations in the models. The results were similar regardless of whether the sibling comparisons included or excluded full-siblings in intact nonstep families (by far the largest subgroup of siblings in this study). In addition, the results did not vary as a function of the residency or amount of visitation between step siblings. Notably, sibling genders and ages also did not affect estimates of sibling similarity. However, the fact that these variables did not influence estimates of sibling similarity does not imply that there are not mean differences in children's behavior problems and prosocial behaviors. As reported previously (Dunn et al., 1998), children in step- and single-mother families, and children whose mothers had experienced multiple partnership changes, were moderately higher in behavior problems and lower in prosocial behaviors compared to children in intact nonstep families. This finding is consistent with the divorce and remarriage literature (Amato & Keith, 1991; Hetherington, 1999).

In conclusion, this test for replication of the findings from twin and adoption studies using the stepfamily genetic design is noteworthy for several reasons. First, the stepfamily design does not confound genetic similarity with sibling similarity in prenatal environments. Second, this large diverse sample of families is more representative of families with sibling children than are twin and adoptive families. Third, although we had only mothers' reports in this study, follow-up research with a selected subgroup of families has shown that mothers' reports are correlated with fathers' and teachers' ratings (Davies et al., 2000). Fourth, the pattern of results is similar to results from adoption studies, in spite of the fact that the stepfamily design (and this study in particular) did not have a restricted range of family environments. The families in the current ALSPAC were very diverse in terms of socioeconomic circumstances, neighborhood safety, exposure to violence in and outside of the home, parental psychopathology, and parenting behaviors (Deater-Deckard & Dunn, 1999). Finally, the sample in the current study included children who had experienced different degrees of family instability. Although this instability clearly influences children's social-emotional adjustment (Hetherington, 1999), it does not appear to influence sibling similarity in mothers' ratings of children's behavior problems and prosocial behaviors. In sum, the stepfamily genetic design provides an additional complementary design for examining gene-environment processes that underlie individual differences in development.

NOTE

1. Intra-class correlations were estimated, this time as partial correlations controlling for the target child's and older sibling's genders (1 = male, 2 = female) as well as the age of the older sibling (all of the target children were 4 years old). For SDQ Total Problems: full-siblings in intact families $r = .37$, full-siblings in other families $r = .30$, half-siblings $r = .14$. For SDQ Prosocial Behavior: full-siblings in intact families $r = .24$, full-siblings in other families $r = .25$, half-siblings $r = .18$. For maternal negativity: full-siblings in intact families $r = .44$, full-siblings in other families $r = .36$, half-siblings $r = .37$. For partner negativity: full-siblings in intact families $r = .52$, full-siblings in other families $r = .48$, half-siblings $r = .36$. Thus, controlling for these variables did not influence estimates of sibling similarity; these partial correlations were very similar to the zero-order intra-class correlations, as reported in Table 2.

REFERENCES

Amato, P. R. (1999). Children of divorced parents as young adults. In E. M. Hetherington (Ed.), *Coping with divorce, single parenting and remarriage: A risk and resiliency perspective* (pp. 47-64). Mahwah, NJ: Erlbaum.

Amato, P. R., & Keith, B. (1991). Parental divorce and the well-being of children: A meta-analysis. *Psychological Bulletin, 110*, 26-46.

Davies, L. C., Dunn, J., O'Connor, T. G., Pickering, K., Deater-Deckard, K., Golding, J., & ALSPAC Study Team (2000). Children's adjustment in stepfamilies, single-parent families and nonstep families: Unpacking the risks and risk processes. Submitted for publication.

Deater-Deckard, K. (2000). Parenting and child behavioral adjustment in early childhood: A quantitative genetic approach to studying family processes. *Child Development, 71*, 468-484.

Deater-Deckard, K., & Dunn, J. (1999). Multiple risks and adjustment in young children growing up in different family settings: A British community study of stepparent, single mother, and nondivorced families. In E. M. Hetherington (Ed.), *Coping with divorce, single parenting and remarriage: A risk and resiliency perspective* (pp. 47-64). Mahwah, NJ: Erlbaum.

Deater-Deckard, K., & Plomin, R. (1999). An adoption study of the etiology of teacher and parent reports of externalizing behavior problems in middle childhood. *Child Development, 70*, 144-154.

Devlin, B., Daniels, M., & Roeder, K. (1997). The heritability of IQ. *Nature, 388*, 468-471.

Dunn, J., Brown, J., & Beardsall, L. (1991). Family talk about feeling states and children's later understanding of others' emotions. *Developmental Psychology, 27*, 448-455.

Dunn, J., Deater-Deckard, K., Pickering, K., O'Connor, T. G., Golding, J., & ALSPAC Study Team (1998). Children's adjustment and prosocial behaviour in step-, single-parent, and non-stepfamily settings: Findings from a community study. *Journal of Child Psychology & Psychiatry, 39*, 1083-1095.

Dunn, J., & Plomin, R. (1990). *Separate lives: Why siblings are so different.* New York: Basic Books.

Eisenberg, N., & Murphy, B. (1995). Parenting and children's moral development. In M. H. Bornstein (Ed.), *Handbook of parenting, Vol. 4: Applied and practical parenting* (pp. 227-257). Mahwah, NJ: Erlbaum.

Eley, T., Deater-Deckard, K., Fombonne, E., Fulker, D. W., DeFries, J. C., & Plomin, R. (1998). An adoption study of depressive symptoms in middle childhood. *Journal of Child Psychology and Psychiatry, 39,* 337-345.

Eley, T., Lichtenstein, P., & Stevenson, J. (1999). Sex differences in the etiology of aggressive and nonaggressive antisocial behavior: Results from two twin studies. *Child Development, 70,* 155-168.

Ge, X., Conger, R. D., Cadoret, R. J., Neiderhiser, J. M., Yates, W., Troughton, E., & Stewart, M. A. (1996). The developmental interface between nature and nurture: A mutual influence model of child antisocial behavior and parent behaviors. *Developmental Psychology, 32,* 574-589.

Golding, J., Pembrey, M., Jones, R., & ALSPAC Study Team. (2001). The Avon Longitudinal Study of Parents and Children; I. Study methodology. *Paediatric & Perinatal Epidemiology, 15,* 74-87.

Goodman, R. (1997). The Strengths and Difficulties Questionnaire: A research note. *Journal of Child Psychology & Psychiatry, 38,* 581-586.

Goodman, R., & Scott, S. (1999). Comparing the Strengths and Difficulties Questionnaire and the Child Behavior Checklist: Is small beautiful? *Journal of Abnormal Child Psychology, 27,* 17-24.

Harris, J. R. (1998). *The nurture assumption.* New York: Free Press.

Haskey, J. (1994). Stepfamilies and stepchildren in Great Britain. *Population Trends, 76,* 17-28.

Hetherington, E. M. (1999). *Coping with divorce, single parenting and remarriage: A risk and resiliency perspective.* Mahwah, NJ: Erlbaum.

Jackson, D. N. (1974). *Personality research form manual* (2nd ed.). Port Huron, MI: Research Psychologists Press.

Kowal, A., & Kramer, L. (1997). Children's understanding of parental differential treatment. *Child Development, 68,* 113-126.

Krevans, J., & Gibbs, J. C. (1996). Parents' use of inductive discipline: Relations to children's empathy and prosocial behavior. *Child Development, 67,* 3263-3277.

Loehlin, J. C. (1992). *Genes and environment in personality development.* Thousand Oaks, CA: Sage.

Mehrabian, A., & Epstein, N. (1972). A measure of emotional empathy. *Journal of Personality, 40,* 525-543.

Miles, D. R., & Carey, G. (1997). Genetic and environmental architecture of human aggression. *Journal of Personality and Social Psychology, 72,* 207-217.

Neale, M. C., & Cardon, L. R. (1992). *Methodology for genetic studies of twins and families.* Dordrecht, The Netherlands: Kluwer Academic Publishers.

O'Connor, T. G., Deater-Deckard, K., Fulker, D. W., Rutter, M., & Plomin, R. (1998). Gene-environment correlations in late childhood and early adolescence. *Developmental Psychology, 34,* 970-981.

O'Connor, T. G., Hetherington, E. M., Reiss, D., & Plomin, R. (1995). A twin-sibling study of observed parent-adolescent interactions. *Child Development, 66*, 812-824.

Patterson, G. R. (1997). Performance models for parenting: A social interactional perspective. In J. Grusec & L. Kuczynski (Eds.), *Parenting and children's internalization of values: A handbook of contemporary theory* (pp. 193-226). New York: Wiley.

Pettit, G. S., Bates, J. E., & Dodge, K. A. (1997). Supportive parenting, ecological context, and children's adjustment: A seven-year longitudinal study. *Child Development, 68*, 908-923.

Pike, A., McGuire, S., Hetherington, E. M., Reiss, D., & Plomin, R. (1996). Family environment and adolescent depressive symptoms and antisocial behavior: A multivariate genetic analysis. *Developmental Psychology, 32*, 590-603.

Plomin, R. (1994a). *Genetics and experience: The interplay between nature and nurture.* Thousand Oaks, CA: Sage.

Plomin, R. (1994b). Genetic research and identification of environmental influence. *Journal of Child Psychology and Psychiatry, 35*, 817-834.

Plomin, R., & DeFries, J. C. (1985). *Origins of individual differences in infancy: The Colorado Adoption Project.* Orlando, FL: Academic Press.

Plomin, R., DeFries, J. C., McClearn, G. E., & Rutter, M. (1997). *Behavioral genetics (3rd ed.).* New York: W. H. Freeman.

Reiss, D., Plomin, R., Hetherington, E. M., Howe, G., Rovine, M., Tyron, A., & Stanley, M. (1994). The separate worlds of teenage siblings: An introduction to the study of nonshared environment and adolescent development. In E. M. Hetherington, D. Reiss, & R. Plomin (Eds.), *Separate social worlds of siblings: Impact of nonshared environment on development* (pp. 63-109). Hillsdale, NJ: Erlbaum.

Rodgers, J. L., Rowe, D. C., & Li, C. (1994). Beyond nature versus nurture: DF analysis of nonshared influences on problem behaviors. *Developmental Psychology, 30*, 374-384.

Rohner, R. (1986). *The warmth dimension: Foundations of parental acceptance-rejection theory.* Beverly Hills, CA: Sage.

Rushton, J. P., Fulker, D. W., Neale, M. C., Nias, D. K. B., & Eysenck, H. (1986). Altruism and aggression: The heritability of individual differences. *Journal of Personality and Social Psychology, 50*, 1192-1198.

Rutter, M., Dunn, J., Plomin, R., Simonoff, E., Pickles, A., Maughan, B., Ormel, J., Meyer, J., & Eaves, L. (1997). Integrating nature and nurture: Implications of person-environment correlations and interactions for developmental psychopathology. *Development and Psychopathology, 9*, 335-364.

Scarr, S., & McCartney, K. (1983). How people make their own environments: A theory of genotype-environment effects. *Child Development, 54*, 424-435.

Sokol, D. K., Moore, C. A., Rose, R. J., Williams, C. J., Reed, T., & Christian, J. C. (1995). Intrapair differences in personality and cognitive ability among young monozygotic twins distinguished by chorion type. *Behavior Genetics, 25*, 457-466.

Stoolmiller, M. (1998). Correcting estimates of shared environmental variance for range restriction in adoption studies using a truncated multivariate normal model. *Behavior Genetics, 28*, 429-441.

Van den Oord, E. J. C. G., & Rowe, D. C. (1997). Continuity and change in children's social maladjustment: A developmental behavior genetic study. *Developmental Psychology, 33*, 319-332.

Zahn-Waxler, C., Robinson, J. L., & Emde, R. N. (1992). The development of empathy in twins. *Developmental Psychology, 28*, 1038-1047.

The Northeast-Northwest Collaborative Adoption Project: Identifying Family Environmental Influences on Cognitive and Social Development

Kirby Deater-Deckard
Stephen A. Petrill
Bessie Wilkerson

SUMMARY. One of the exciting new directions in family research is the examination of shared and nonshared environmental processes un-

Kirby Deater-Deckard is affiliated with the University of Oregon. Stephen A. Petrill is affiliated with the Pennsylvania State University. Bessie Wilkerson is affiliated with Wesleyan University (CT).

Address correspondence to Kirby Deater-Deckard, Department of Psychology, 1227 University of Oregon, Eugene, OR 97403-1227 (E-mail: kirbydd@darkwing.uoregon.edu).

The authors thank the parents, children, and adoption agency personnel for their participation. The authors also wish to thank their research teams at the University of Oregon and Wesleyan University for their help with data collection and management, and Leslie Leve and John Reid at the Oregon Social Learning Center for their help in the initial recruitment of participants in the Northwest Adoption Project, funded by NIMH grant P50 MH46690 to John Reid.

This research was supported by grants to the first and second authors from the National Science Foundation (BCS9907811, BCS9907860) and the Society for the Psychological Study of Social Issues. The authors contributed equally to the writing of this manuscript. Portions of this manuscript were submitted as grant applications to the National Science Foundation (February 1999) and the National Institutes of Health (July 2000).

[Haworth co-indexing entry note]: "The Northeast-Northwest Collaborative Adoption Project: Identifying Family Environmental Influences on Cognitive and Social Development." Deater-Deckard, Kirby, Stephen A. Petrill, and Bessie Wilkerson. Co-published simultaneously in *Marriage & Family Review* (The Haworth Press, Inc.) Vol. 33, No. 2/3, 2001, pp. 157-178; and: *Gene-Environment Processes in Social Behaviors and Relationships* (ed: Kirby Deater-Deckard, and Stephen A. Petrill) The Haworth Press, Inc., 2001, pp. 157-178. Single or multiple copies of this article are available for a fee from The Haworth Document Delivery Service [1-800-HAWORTH, 9:00 a.m. - 5:00 p.m. (EST). E-mail address: getinfo@haworthpressinc.com].

157

derlying the learning of cognitive, literacy, and social-behavioral skills. The adoption design is a powerful method for testing theories regarding environmental mechanisms. In this paper, we describe the Northeast-Northwest Collaborative Adoption Project (N2CAP), an ongoing adoption study in which we are examining the role of parenting environments generally, and parent-child relationships more specifically, in the development of cognitive, reading and social-emotional outcomes in childhood. The goal of this study is to identify environmental influences that impact children's cognitive and social-emotional development using a genetically sensitive design, and to further the development of models of genotype-environment processes. *[Article copies available for a fee from The Haworth Document Delivery Service: 1-800-HAWORTH. E-mail address: <getinfo@haworthpressinc.com> Website: <http://www.HaworthPress. com> © 2001 by The Haworth Press, Inc. All rights reserved.]*

KEYWORDS. Adoption, family environment, social development, literacy

The acquisition of cognitive and social-emotional skills is an inherently social process. These social interactions typically occur within the context of an immediate and extended family network that is itself embedded within a community and a larger cultural group. Prior to and following the child's entrance to more formal schooling, the family environment provides many opportunities for acquiring skills and knowledge that are appropriate in that child's culture (Bronfenbrenner, 1981; Rogoff, 1991). At the same time, genetic influences in skills and behaviors are known to be present (Plomin, 1994). One of the exciting new directions in developmental and family psychology is the examination of the ways in which these environmental and genetic influences work together in the development of the "whole" child–that is, understanding individual differences in children's cognitive as well as social-emotional development.

Research of this sort requires an integration of theory and methods across a variety of approaches. This paper describes one such research initiative, the Northeast-Northwest Collaborative Adoption Project (N2CAP). This is an ongoing study of cognitive and social development that employs two behavioral genetic designs–the adoptive parent-offspring design, and the adoptive sibling design–that we recently initiated in order to test theories regarding family environmental mechanisms.

We begin by providing a brief overview of existing research examining links between parenting environments, cognitive development, literacy, and social-emotional adjustment. This is followed by a discussion of some of the advantages of using a behavior genetic design to answer questions about environmental processes. Finally, we provide a description of the methods and aims of the study, along with some preliminary results.

COGNITIVE AND SOCIAL DEVELOPMENT:
ENVIRONMENTAL MECHANISMS

There is considerable overlap between children's healthy and maladaptive development in the cognitive (including literacy) and social-emotional realms. We know that children's behavioral problems, emotional problems, and difficulties in cognitive performance, reading, and academic achievement co-occur (for example, see Hinshaw, 1992; Silva, Williams, & McGee, 1987; Tomblin, Zhang, Buckwalter, & Catts, 2000). It is likely that there is overlap in the types of family environmental influences operating on this wide array of child outcomes.

First, consider the research on family environmental influences on cognitive development–much of this work has focused on intelligence and specific cognitive abilities. These studies may be divided into three broad categories (see Brody, 1992, for a review). Attributes of the *biological environment*–illnesses, prenatal effects, postnatal events, nutrition, high lead levels–are moderately correlated with measures of general cognitive development in children. Other studies have attempted to examine more psychological and educational aspects of the environment that relate to cognitive outcomes. The *cognitive correlates* approach (see Gottfried, 1984; Hart & Risley, 1995) examines links between cognitive outcomes and measures of stimulation and enrichment such as the HOME Inventory (Bradley & Caldwell, 1976, 1980; Caldwell & Bradley, 1978), yielding an average correlation with IQ of .30. Finally, the *intervention* approach (see Spitz, 1986) has demonstrated that one can raise the average cognitive test performance of economically disadvantaged groups of children through intensive individualized mentoring. For example, the Abecedarian Project (Ramey & Campbell, 1991) involved an extensive daycare program for children judged to be at risk for academic failure. Average age at entry was 4.4 months. The Abecedarian Project endeavored to involve families in the preschool program and performed a support service to improve communication between parents and teachers once the children reached elementary school. Effective in the short-term, Campbell and Ramey (1994) also reexamined these children when they were 12, and found that a 1/3 standard deviation gain in IQ had remained over the intervening years.

Although much of this research on family environments has focused on general cognitive ability, researchers have also examined environmental influences on related outcomes such as literacy. Literacy is linked fundamentally to general cognitive ability, but also to children's educational outcomes, their eventual occupational success, and their concurrent and subsequent social-emotional adjustment. Many studies have focused on identifying environmental influences on literacy skills. A large number of these have shown a modest correlation ($r = .28$) between shared reading (i.e., parents reading to their preschool children) and later reading success, although this effect disappears in early grades once socioeconomic factors and children's previous cognitive and linguistic skills are controlled (Scarborough & Dobrich, 1994; Bus, van Ijzendoorn & Pellegrini, 1995). Similarly, availability of books and other literacy materials in the home correlates .27 with later reading success (Scarborough, 1998). However, what matters may be how parents and teachers read to children, and not how much. Although studies have not shown evidence of substantial correlations to support this hypothesis (Scarborough & Dobrich, 1994), interventions teaching parents and teachers how to share a book have yielded positive influences on oral language development (Mason, 1992; Moon & Wells; 1979, Lonigan, 1993; Whitehurst et al., 1988). Additionally, parents' positive attitudes toward and expectations about reading and school are stronger environmental predictors of reading precocity than actual book reading time (Briggs & Elkind, 1977; Dunn, 1981; Scarborough & Dobrich, 1994).

Second, researchers have recognized that we cannot fully understand children's cognitive and literacy skills without also examining children's social, behavioral and emotional development—good mental health is an essential element of desirable cognitive and literacy outcomes. Decades of research have shown that there are links between parenting and children's emotion regulation and social competence (e.g., positive peer relationships, self-control), with parental warmth and firm control deriving optimal outcomes for most children (Maccoby & Martin, 1983). Similarly, harsh coercive parenting practices have been implicated in the development and maintenance of psychopathology in childhood (e.g., behavior problems and emotional disturbance) that are not only deleterious in their own right but are linked to problems in learning and failure in school (Hinshaw, 1992). More recently, these same components of the parenting environment–warmth/rejection and extent and type of control–have been implicated in social-cognitive domains of development, including social information processing deficits (Dodge, Bates, & Pettit, 1990) and understanding of other people's thoughts and beliefs (Hughes, Deater- Deckard, & Cutting, 1999).

Taken together, the research on children's cognitive and literacy skills and social-emotional development suggests that the family environment–and parenting in particular–is critical to the processes underlying the establishment and maintenance of individual differences in these related outcomes. In the next section of the paper, we consider one of the ways in which behavior genetics researchers are able to disentangle some of these family environmental processes from genetic influences.

WHAT WE CAN LEARN BY STUDYING ADOPTIVE FAMILIES

Adoptive family research designs are well suited for studying family environmental processes. Although at first glance it may seem difficult to identify and study this population of families, this method is feasible, largely because of the increasing numbers of international adoptions that are taking place in many countries. Already, studies of children adopted from deprived environments early in life and placed in enriching adoptive homes have shown that many of these children show massive gains in cognitive ability and physical stature, particularly when they have been adopted as a result of war or institutionalization in their countries of origin (Tizard, 1991). For example, research in Britain on the development of orphaned Romanian children has shown that being placed in a caring adoptive home has a dramatic effect on developmental outcomes (Rutter, 1998). Many of these children are able to "catch up," with the amelioration of early deprivation being most dramatic for children who are placed in their adoptive homes at younger ages. This is just one example of the utility of the adoptive family design, with respect to identifying environmental mechanisms.

The large literature regarding children's rearing environments and their cognitive, literacy, and social-emotional development provides us with a strong base from which we can examine the nature of environmental mechanisms by using an adoptive family design. We will focus on at least two issues that require further research and that provide a core set of questions to be addressed in this study–the lack of independence of genetic and environmental influences, and the examination of "child specific" as well as global environments within families.

Genetic and Environmental Factors Are Not Independent

The first issue is that genetic and environmental influences on children's developmental outcomes are not independent. By examining genetically related parents and children who also share their environments, genetic and environmental influences are confounded and cannot be distinguished (Scarr, 1992). This is a concern because twin and adoption studies have shown that genetic influences are known to play a role in reading outcome, cognitive abilities, and social-emotional development and psychopathology (Bouchard & McGue, 1981; Chipuer, Rovine, & Plomin, 1990; DeFries, Stevenson, & Gillis, & Wadsworth, 1991; Plomin, 1994; Vogler & DeFries, 1986). More recently, researchers have begun to identify DNA markers that are associated with reading disability (Cardon et al., 1994) and associated phonological deficits (Grigorenko et al., 1997), high vs. average IQ scores (Chorney et al., 1998), and personality and psychopathology (e.g., novelty seeking, see Benjamin et al., 1996; Ebstein et al., 1995).

Because parents provide children with their genes as well as their rearing environments we are not able to assume that the causal pathways involving "reading" or "social" environments that are provided to children by their parents are free from child genetic influences. There is evidence from behavioral genetic studies that child genetic influences operate, in part, through transactions with the environment–through gene-environment correlations and interactions (Deater-Deckard, 2000; Ge, Conger, Cadoret, Neiderhiser et al., 1996; O'Connor, Deater-Deckard, Fulker, Rutter, & Plomin, 1998; O'Connor, Hetherington, Reiss, & Plomin, 1995; Pike, McGuire, Hetherington, Reiss, & Plomin, 1996). For example, sibling differences in reading skills that may include an underlying genetic component can be reinforced as these children seek out different experiences (e.g., the child who is more adept at reading may find reading more rewarding, thereby reading much more frequently than his or her sibling), or elicit different reactions from their parents and peers (e.g., parents read more frequently with the child who is already more adept and interested in reading). Although the bulk of the earlier behavioral genetic literature did not identify precisely which environmental factors might be implicated in these gene-environment transactions, this is changing rapidly. Increasingly common are studies that integrate traditional family environment research methods (e.g., observations of parenting behavior) and traditional behavior genetic methods (e.g., examining genetically unrelated family members). Recent examples include the Nonshared Environment in Adolescent Development study (NEAD: Reiss, Neiderhiser, Hetherington, & Plomin, 2000) and the ongoing longitudinal Colorado Adoption Project (DeFries, Plomin, & Fulker, 1994), both of which have served as models for the current adoption study.

Child-Specific Family Environmental Processes

A second issue concerns the way in which the environment is typically conceptualized and assessed. Most previous studies have examined home environment variables that are known to vary between different families–for example, socioeconomic factors, maternal education, parental occupation, parenting practices, and the number of books in the home. Although these global environmental variables inform our understanding of development, what is needed are studies that also examine environmental variables at the level of selected individuals within each family–for instance, time spent in shared reading between a particular parent and a particular child within each family.

An assumption in much of the research has been that these environmental influences are most likely operating as part of "shared environmental processes"–that is, they create sibling and parent-offspring similarity among co-residing family members for reasons of shared experience rather than shared genetic influences. Thus, one of our principal aims is to test for the presence of such shared environmental processes by examining similarity among genetically unrelated adoptive family members. By testing for adoptive parent-child and sibling similarity in these developmental outcomes, and then striving to identify those aspects of the family and parenting environment that account for this similarity among genetically unrelated family members, we can test directly theories regarding the role of these shared environmental processes.

At the same time, the evidence from behavior genetic and socialization studies together points to "nonshared environmental processes" as being important–that is, environmental processes that lead to differentiation of co-residing family members. Indeed, most behavior genetic studies suggest that much if not most of the environmental variance in cognitive and social-emotional outcomes is of this "nonshared" variety (Plomin & Daniels, 1987). Examination of these nonshared environmental processes has become more common in the past decade (see

Reiss et al., 2000), although much remains to be learned about which specific environmental factors are involved, and how these nonshared environmental processes actually operate (McGuire, this volume; Turkheimer & Waldron, 2000). The current study's design will not allow us to distinguish these nonshared environmental processes from sibling differences that arise as a result of genetic differences. However, we are still hopeful that by examining child-specific environments within each family, we may be able to identify potential candidate nonshared processes that can be tested using other genetic designs.

The research that we describe next is multidisciplinary (incorporating methods from developmental psychology and behavioral genetics) and multi-method (utilizing questionnaires, interviews, standardized testing procedures, real-time and videotaped observations). A model for examining environmental processes will be tested by identifying shared environmental processes that statistically predict cognitive and social developmental outcomes. This examination of family environmental influences will include intensive assessments of children's interactions with their parents and siblings, the influence of family resources including literacy materials and socioeconomic resources, as well as an examination of demographic and cultural characteristics within the sample of multiethnic adoptive families. By delineating more precise environmental mechanisms, the research will contribute to the development and targeting of preventative and ameliorative models that can promote children's healthy and adaptive cognitive and social-emotional outcomes.

THE NORTHEAST-NORTHWEST COLLABORATIVE ADOPTION PROJECT (N2CAP)

Next, we turn to a description of the Northeast-Northwest Collaborative Adoption Project (N2CAP). Because adoptive family members are genetically unrelated, any statistically significant association in behavior and skills found between adoptive parents and their adopted children, or between unrelated siblings, is a direct estimate of shared environmental influences. Once similarity among adoptive family members is detected, we can examine various aspects of the family environment that may help account for this similarity. For example, if there is an effect of "number of books in home" upon sibling similarity in adopted children's reading outcomes, we will be able to strongly conclude that this is a shared environmental influence that cannot be attributed to shared genes for reading and cognitive ability among related family members. The long-term goal of this project is to assess changes in these environmental processes and outcomes throughout childhood and adolescence within a longitudinal design. The methodology of the N2CAP study is described below.

Methods

The N2CAP includes two studies, a large national survey and a smaller regional in-depth assessment study.

Sample for the Survey Study. The first study is a national survey of adoptive families and their children. The identification of potential participants and invitation to participate occurs in several stages. First, research staff identify adoption agencies and lawyers in the New England and Pacific Northwest regions of the U.S. and contact them by mail to ask if they are interested in collaborating in the study. These letters to the agencies are followed by a telephone call by research staff. Agencies that agree then mail, on our behalf, a letter of invitation and brochure to families that they have worked with in the past. Included in the letter to families is a response form that they return to us if they are interested in participating.

At this time, we have heard back from over 2000 parents who may be interested in participating in the survey. Of these, approximately 1600 have begun providing demographic information, and about 400 have completed all or part of the survey. We hope to collect survey data from as many of these interested families as is possible. The majority of these participants will include families who have at least one adopted child, with most children ranging in age from 0-10 years. Ultimately, our goal is to establish an ongoing longitudinal survey study of adopted children, their siblings, and their adoptive parents. Ideally, we will eventually be able to invite these families to participate in other studies involving in-home and laboratory assessments.

Procedures for the Survey Study. Parents who return a form expressing their interest in the study are then mailed a brief demographics questionnaire. This questionnaire includes information regarding household structure, parental education and occupation, marital status, and details regarding all of their children (i.e., adopted, biological, stepchild, international or domestic adoption, country of origin, ethnicity, birth date, adoption date). Parents also provide signed informed consent at this time. The questionnaire is returned by mail to research staff.

Next, participating parents complete the survey materials. Parents can complete the survey through the mail or by using the Internet. Those parents who use the Internet to complete the survey are provided a unique PIN number to ensure confidentiality.

Measures for the Survey Study. A list of the measures employed in N2CAP (including citations) is provided in Table 1.

Academic Achievement. For those children who are old enough, parents provide information pertaining to the child's current academic

TABLE 1. Measures Used in the Survey and In-Depth Studies of the Northeast-Northwest Adoption Project

Child cognitive/reading/achievement
Parent Report of Academic Achievement (developed for current study)[1]
Stanford-Binet Intelligence Scale (Thorndike, Hagen, & Sattler, 1986)[2]
Phonological Awareness Test (Robertson & Salter, 1997)[2]
Woodcock Reading Mastery Test-Revised (Woodcock, 1987)[2]
Peabody Picture Vocabulary Test (Dunn, Dunn, & Dunn, 1997)[2]
Parent cognitive/reading/achievement
Stanford-Binet Intelligence Scale (Thorndike, Hagen, & Sattler, 1986)[2]
Woocock-Johnson Tests of Achievement (Woodcock & Johnson, 1989)[2]
Child social-emotional
Child Behavior Checklist (Achenbach, 1991)[1]
Strengths & Difficulties Questionnaire (Goodman, 1997)[1]
Colorado Childhood Temperament Inventory (Rowe & Plomin, 1977)[1]
Infant Behavior Record (Bayley, 1993)[2]
Parent-Child Interaction System (Deater-Deckard, Pylas, & Petrill, 1997)[2]
Home environment
Experience of Adoption Questionnaire (developed for current study)[1]
International Adoption Questionnaire (adapted from Tessler, Gamache, & Liu, 1996)[1]
Educational Attitude Scale (Rescorla, Hyson, Hirsh-Pasek, & Cone, 1990)[1]
The Educational Environments Questionnaire (developed for current study)[1]
Parental Modernity Scale (Schaefer & Edgerton, 1985)[1]
Parent Feelings Questionnaire (Deater-Deckard, 1996)[1]
Discipline Interview (adapted from Deater-Deckard, 2000)[2]
Parent-Child Interaction System (Deater-Deckard, Pylas, & Petrill, 1997)[2]
Language Coding System (Petrill, Bermont, & Kruming, 1998)[2]
Post-Visit Inventory (adapted from Dodge, Bates, & Pettit, 1986)[2]

[1]Administered as part of national survey study.
[2]Administered as part of in-depth study.

achievement, including school grades and, when available and known, standardized scholastic aptitude test scores.

Child Social-Emotional Development. Parents complete ratings of the child's behavioral (e.g., aggression, conduct problems) and emotional adjustment (e.g., anxiety, depressive symptoms) using the Child Behavior Checklist and Strengths and Difficulties Questionnaire. Parents also complete ratings of the child's temperament using the Colorado Childhood Temperament Inventory, which includes scales representing sociability, shyness, emotionality, activity level, and persistence.

Home Environment. Parents provide information regarding their experiences in the preparation and culmination of the adoption process using the Experience of Adoption Questionnaire, designed for this study. When applicable, they also complete the International Adoption Questionnaire, designed for this study and adapted from a survey instrument developed by Tessler that measures parents' attitudes regarding children's multicultural heritage and bicultural socialization. In addition, they report their attitudes about the importance of academic, athletic, artistic, and social experiences for young children using the Educational Attitude Scale. Parents also complete a questionnaire designed for this study pertaining to the provision of educationally related environments for each child (e.g., how many books/magazines do you read in a month, how many books/magazines does your child read in a month).

Parents also report on their childrearing attitudes, using the Parental Modernity Scale. This yields a traditionalism scale that is a correlate of authoritarian parenting as well as children's cognitive and social-emotional outcomes. In addition, parents provide ratings of their own negative and positive feelings toward each child using the Parent Feelings Questionnaire.

Sample for the In-Depth Study. The second study is an in-depth investigation involving a visit to the family's home. For this second study, we plan to include 300 families from the national survey sample who reside in the New England and Pacific Northwest regions of the United States. The purpose of the second study is to gather data using multiple procedures and data sources, to operate as a complement to the larger survey study. At this time, we have completed about 150 of these home visits.

Procedures for the In-Depth Study

In the in-depth study of 300 families, we conduct a 2- to 3-hour visit to each family's home. Two researchers administer a battery of social and cognitive assessments with up to two children and one or two parents in each household. In addition, the researchers videotape parent-child and sibling interaction using structured tasks. In the first structured task, the parent and child are asked to draw some pictures together using an Etch-A-Sketch toy. In the second structured task designed for this study, the parent and child play a tilting maze game, where a marble must be maneuvered through a maze while avoiding holes and dead ends. Both tasks require cooperation between parent and child, and elicit a wide variety of responses from the parents and children. These videotapes are subsequently coded to assess the language environment, child and parent behavior, and various aspects of the parent-child relationship. Finally, about nine months after the home visit, a follow-up phone call is conducted.

Measures for the In-Depth Study

General Cognitive Ability and Reading Related Skills. Parents' and children's specific cognitive abilities are assessed using a short form of the Stanford-Binet Intelligence Scale. Assessments of children's and parents' reading-related cognitive skills are also conducted.

Phonological Awareness. Phonological awareness (PA) is central to understanding the development of reading. Phonological awareness prior to formal reading instruction predicts about one-fifth of the variance in later reading ability scores (Scarborough, 1998). PA is assessed using the Phonological Awareness Test. This test provides standardized scores for rhyming (discrimination and production), phoneme analysis (segmentation and isolation), grapheme-phoneme knowledge, and decoding (pseudo-word reading), as well as a total score. This will allow us to distinguish phonological sensitivity (most notably rhyming) from phonological knowledge.

Letter Knowledge. Letter knowledge (LK) is the strongest predictor of later reading success, accounting for a quarter of the variance (Scarborough, 1998). To assess LK, children complete the Letter Identification subtest of the Woodcock-Johnson Reading Mastery Test-Revised (WRMT-R).

Reading Outcomes. Reading outcomes include measures of word recognition, non-word reading, and reading comprehension for both children and their parents. The Word Identification and Word Attack subtests are being employed from the WRMT-R for children, and the Woodcock-Johnson Achievement Test for parents. Finally, children complete the Peabody Picture Vocabulary Test as a measure of the complexity of receptive vocabulary.

Child Social-Emotional Development. The researcher conducting the assessment of the child completes the Infant Behavior Record, a measure of temperament that includes scales for positive mood, orientation toward examiner, motor control, and a total score. Following data collection, different observers complete global ratings of child emotion and behavior from videotaped observations collected during the home visit, using the Parent-Child Interaction System (PARCHISY), which includes scales for: noncompliance, negative and positive affect, autonomy, persistence or "on task" behavior, responsiveness to the parent, activity level, and frequency of verbalizations. A more detailed examination of sibling differences in language-based aspects from the videotaped interactions will also be examined using the Language Coding System.

Home Environment. General as well as child-specific measures of various aspects of the family and home environment are assessed using interviewers' and observers' ratings. These measures have been developed from research on families with twins and adoptive siblings, and have been demonstrated to be correlated with children's cognitive and social-emotional outcomes (Deater-Deckard, 2000; Hughes et al., 1999). Following the home visit, researchers conduct an adapted version of the Post-visit Inventory. Using this questionnaire, the researcher records structural characteristics of the home (i.e., type of dwelling, crowding, toys and books for children, presence of any dangers like heavy traffic or unhealthy conditions), as well as their impressions of maternal and paternal warmth and negativity.

During the follow-up phone call, parents complete a brief interview regarding the types and frequency of various discipline strategies they use with each child. From this, the interviewer provides not only detailed data regarding the discipline methods used by each parent (ranging from reasoning to spanking to "time outs"), but also completes a global impression regarding the overall harshness of discipline used (1 = nonrestrictive, mostly positive guidance, to 5 = severe, strict, usually physical punishment). This interview and global rating is based on a clinical interview used in large community studies of preschool and school-age children (Dodge, Bates & Pettit, 1990).

Observers also complete global ratings of child-specific parenting behavior from the videotaped parent-child interaction recorded during the home visit, again using the PARCHISY. Coded parenting behaviors include warmth and negativity, positive and negative control, persistence, responsiveness to child, and frequency of verbalizations. Coded parent-child dyadic interaction behaviors include conflict, cooperation, and emotional reciprocity (i.e., shared positive affect). In addition, the Language Coding System is used to assess parent-based linguistic aspects of the videotaped parent-child interactions. These include complexity of parental utterances, use (or lack of use) of child-directed speech, and types of language used by parents to direct the completion of the structured tasks.

Aims and Hypotheses

In this final section, we describe our aims and hypotheses, present some very preliminary results, and highlight planned analyses. We will address four major aims.

Aim 1: Examining parent-offspring and sibling similarity in cognitive and social-emotional developmental outcomes. Among biologically unrelated siblings who are reared in the same home environments and schools, sibling similarity in cognitive, literacy and behavioral outcomes provides the only *direct* and most reliable estimate of shared environmental variance–that is, environments that lead to sibling similarity. Similarly, among adoptive parents and their genetically unrelated children, parent-child similarity in these cognitive and literacy skills provides another direct estimate of shared environmental variance. This is a crucial first step toward identifying these shared environmental mechanisms–finding evidence of family member similarity in these outcomes among genetically unrelated family members.

The examination of parents and their adopted children, and the examination of unrelated siblings, may yield different results. It seems reasonable to expect adoptive family member similarity to be highest among siblings because the children are usually close in age, whereas parent-child similarity may be lower because of large age differences. Even though we are using the same instruments to assess parents' and children's cognitive and literacy skills, there may be developmental discontinuities in these constructs that result in lower estimates of parent-child similarity.

Aim 2: Examining mediation of distal family environmental factors (e.g., socioeconomic status, parental education) on children's cognitive and social-emotional outcomes via proximal child-specific family environment factors (e.g., parent-child relationship processes, access to literacy materials). Identifying shared environmental sources of variance (aim 1) is a first step, but this alone does not tell us how these environmental mechanisms operate. To do this, we must test potential environmental processes. One goal of these analyses will be to examine the extent to which aspects of the home environment are global and general, as opposed to being divided into numerous independent components. Additionally, we will examine the extent to which proximal environmental measures may be predicted by a set of more distal measures. For example, are behavioral and cognitive outcomes predicted by distal aspects of the home environment (e.g., SES) or are they predicted by specific behavioral and/or cognitive environmental influences (e.g., parental warmth, shared reading)? Although it is premature to posit specific models, structural equation modeling approaches will be implemented to explore the relationships between environmental measures and cognitive outcomes.

N2CAP is in its infancy, but we have conducted some preliminary data analyses based on the first 26 participating families. We include these results here to provide examples of some of the ways in which we will analyze these data once the assessments are complete.

One area of environmental influence on children's development that has received attention is parents' childrearing attitudes. We examined the associations between parental attitudes and the cognitive measures in this preliminary sample. Variations in fathers' educational and social attitudes were moderately to substantially associated with their adopted children's cognitive and literacy scores. For example, fathers' educational and social attitudes, as measured using the Educational Attitudes Questionnaire, were moderately to substantially associated ($r = -.5$ to $-.7$ range) with their adopted children's cognitive and literacy scores (e.g., Stanford-Binet memory, Woodcock-Johnson word identification and word attack).

Another aspect of the parenting environment that is crucial is parental warmth and negativity. Parents' self-reported positive and negative feelings toward each child (measured using the Parent Feelings Questionnaire) were associated with their adopted children's cognitive development. For example, the correlations for mothers' self-reported positivity and negativity and their children's Stanford-Binet scores are shown in Table 2.

These and a variety of other indicators of the parenting environment (see Table 1) are anticipated to emerge as candidates for examining environmental mediation of these and other developmental outcomes, the next step and major thrust of the planned analyses for this study.

Aim 3: Examining the shared environmental mediation of family member similarity via these family environmental factors. With shared environmental variance (aim 1) and likely environmental processes having been delineated (aim 2), next we need to apply what we learn

TABLE 2. Maternal Positivity and Negativity in Relation to Stanford-Binet Scores

	Verbal	Abstract	Memory	Composite
Positivity Scale A	.45*	.24	.52*	.33
Positivity Scale B	.23	.52*	.60**	.53*
Negativity Scale A	−.28	−.60**	−.44	−.58**
Negativity Scale B	−.23	−.34	−.56*	−.55*

Note: n = 18 to 20; * p < .05, ** p < .01. Positivity and negativity were measured using two scales from the Parent Feelings Questionnaire; scale A was rated on a 5-point Likert-type scale, and scale B was rated on a 10-point ordinal scale.

about these environmental factors to see if they in fact account for the shared environmental variance that we find in these cognitive and social-emotional outcomes. To do this, we will examine whether these environmental processes mediate the association between parents' and adopted children's cognitive and social-emotional outcomes, as well as the association between unrelated adoptive siblings' outcomes. The essential test is whether or not we are able to explain why genetically unrelated family members are similar to one another in these cognitive skills and social behaviors, using those aspects of the family environment that we have measured.

We will employ a modeling strategy adapted from Pike et al. (1996). In their original analysis, Pike et al. decomposed the correlation between measures of the environment and behavioral outcomes into genetic, shared environmental, and nonshared environmental components. Although this full decomposition is not possible using this research design, it is possible to decompose a correlation between a measured environment and an outcome into shared environmental (C) and nonshared or "child specific" (SPEC) genetic + environmental sources of variance (see Figure 1).

According to this model, there are sources of C and SPEC that yield the correlation between child environment and outcome. A significant general C suggests that the correlation between measured environment and outcome is due to shared family influences. A significant general SPEC suggests that the correlation between measured environment and outcome is due to child specific processes that may have either or both environmental and genetic origins. Additionally, there are unique

FIGURE 1. Sibling mediation model. Child 1 and Child 2 refer to genetically unrelated siblings who are being reared together in the adoptive home. "C" refers to that portion of the covariation between the identified environmental measure and child outcome that is accounted for by common or shared environmental processes. "SPEC" refers to that portion of the covariation that is accounted for by nonshared genetic + environmental processes. "c" and "spec" refer to the remaining variation that is not accounted for by "C" and "SPEC."

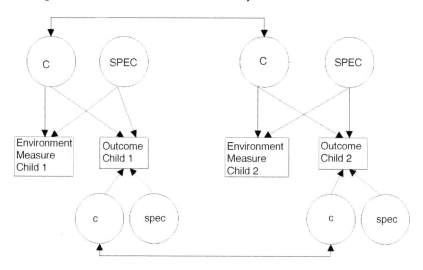

sources of "c" and "spec." A significant unique "c" estimate suggests that independent shared environmental variance affects outcome separately from shared environmental variance in the environmental measure. A significant unique "spec" suggests that these outcomes are uncorrelated among these genetically unrelated family members.

Thus, an environmental measure can be decomposed into shared and child-specific components. For example, parents use some general discipline strategies that they apply to both children in a family (e.g., some parents use spanking, whereas others never use spanking). Thus, the general tenor of discipline differs between families, and can be predictive of children's outcomes in a way that would be characterized as a shared environmental influence if it leads to sibling similarity. At the same time, despite these variations in parental discipline between families, it may also be true that child-specific variability in discipline within the same family may be associated with differences in sibling outcomes (e.g., one child is treated much more severely than her sibling); this would suggest that child-specific processes are operating as well.

It is important to note that this design will not allow us to specify whether these child-specific processes are genetically or environmentally mediated. That is, the design of N2CAP is very powerful for identifying shared environmental processes, but cannot be used to define nonshared environmental processes. Nonetheless, examining this "spec" component of variance in these environmental measures and child outcomes may lead to the identification of potential "candidate" nonshared environmental processes–candidates that can be tested using genetic designs that are able to separate genetic and nonshared environmental components.

CONCLUSIONS

The environments that parents provide play a central role in the cognitive, social-emotional, and behavioral development of children. Early behaviorist theories suggested that children were "blank slates" whose cognitive, social-emotional, and behavioral development was shaped almost entirely by parental influence. More recently, theory and research have moved away from this unidirectional view and toward the idea that children and parents influence one another, actively constructing together the experiences that are relevant to social, behavioral, and cognitive development (Bugental & Goodnow, 1998). Children are not passive recipients of their environments. Instead, the home environment is the result of a dynamic process involving parent and child attributes.

Behavioral genetic evidence suggests that both genetic and environmental factors contribute to cognitive, social-emotional, and behavioral development. The focus of research is no longer on whether something shows genetic or environmental influence, but how genes and environments produce a rich human diversity in cognitive and social-emotional outcomes (Plomin, 1994). In order to study these mechanisms, genetically sensitive designs are required. Adopted children are genetically unrelated to their adoptive parents, and in most adoptive families, are also unrelated to one another. Thus, any associations found between adopted children's developmental outcomes and their adoptive parents' characteristics, or between adoptive siblings' outcomes, cannot be attributed to shared genes but instead implicates shared environmental processes. Furthermore, unlike other behavior genetic methods that estimate these shared environmental effects indirectly (e.g., twin designs), adoption studies offer a direct estimate of shared environmental processes in development. Finally, this research will contribute to our knowledge about those aspects of the family environment and the adoption process that promote optimal outcomes for adopted children and their parents. This knowledge may one day be applied in a way that promotes adaptation for these children, their siblings, and parents.

REFERENCES

Achenbach, T. M. (1991). *Integrative guide for the 1991 CBCL/4-18, YSR, and TRF profiles.* Burlington, VT: University of Vermont Department of Psychiatry.

Bayley, N. (1993). *Manual for the Bayley Scales of Infant Development* (2nd Edition). San Antonio, TX: Psychological Corporation.

Benjamin, J., Li, L., Patterson, C., Greenberg, B. D., Murphy, D. L., & Hamer, D. H. (1996). Population and familial association between the D4 Dopamine Receptor gene and measures of novelty seeking. *Nature Genetics, 12,* 81-84.

Bouchard, T. J. Jr., & McGue, M. (1981). Familial studies of intelligence: A review. *Science, 212,* 1055-1059.

Bradley, R., & Caldwell, B. (1976). Early home environment and changes in mental test performance from 6 to 36 months. *Developmental Psychology, 12,* 93-97.

Bradley, R., & Caldwell, B. (1980). Home environment, cognitive competence, and IQ among males and females. *Child Development, 51,* 1140-1148.

Briggs, C., & Elkind, D. (1977). Characteristics of early readers. *Perceptual and Motor Skills, 44,* 1231-1237.

Brody, N. (1992). *Intelligence* (2nd ed.). San Diego, CA: Academic Press.

Bronfenbrenner, U. (1981). *Ecology of human development: Experiments by nature and design.* Cambridge, MA: Harvard University Press.

Bugental, D. B., & Goodnow, J. J. (1998). Socialization processes. Chapter in N. Eisenberg (Volume Ed.), *Handbook of Child Psychology, Vol. 3* (pp. 389-462). New York: Wiley.

Bus, A. G., van Ijzendoorn, M. H., & Pellegrini, A. (1995). Joint book reading makes for success in learning to read: A meta-analysis on intergenerational transmission of literacy. *Review of Educational Research, 65,* 1-21.

Caldwell, B., & Bradley, R. (1978). *Home observation for measurement of the environment.* Little Rock: University of Arkansas.

Campbell, F. A., & Ramey, C. T. (1994). Effects of early intervention on intellectual and academic achievement: A follow-up study of children from low income families. *Child Development, 65,* 684-698.

Cardon, L. R., Smith, S. D., Fulker, D. W., Kimberling, W. J., Pennington, B. F., & DeFries, J. C. (1994). Quantitative trait locus for reading disability on Chromosome 6. *Science, 266,* 921-923.

Chipuer, H. M., Rovine, M. J., & Plomin, R. (1990). LISREL modeling: Genetic and environmental influences on IQ revisited. *Intelligence, 14,* 11-29.

Chorney, M. J., Chorney, K., Seese, N., Owen, M. J., Daniels, J., McGuffin, P., Thompson, L. A., Detterman, D. K., Benbow, C., Labinsky, D., Eley, T., & Plomin, R. (1998). A quantitative trait locus associated with general cognitive ability in children. *Psychological Science, 9,* 159-166.

Deater-Deckard, K. (1996). *The Parent Feelings Questionnaire.* London, UK: Institute of Psychiatry.

Deater-Deckard, K., Pylas, M. V., & Petrill, S. A. (1997). *The Parent-child interaction system (PARCHISY).* London, UK: Institute of Psychiatry.

Deater-Deckard, K. (2000). Parenting and child behavioral adjustment in early childhood: A quantitative genetic approach to studying family processes and child development. *Child Development, 71,* 468-484.

DeFries, J. C., Plomin, R., & Fulker, D. W. (1994). *Nature and nurture during middle childhood.* Cambridge, MA: Blackwell.

DeFries, J. C., Stevenson, J., Gillis, J. J., & Wadsworth, S. J. (1991). Genetic etiology of reading deficits in the Colorado and London Twin Studies of reading disability. *Reading and Writing, 3,* 271-283.

Dodge, K. A., Bates, J. E., & Pettit, G. S. (1986). Development of aggressive behavior. Research grant proposal submitted to National Institutes of Mental Health.

Dodge, K. A., Bates, J. E., & Pettit, G. S. (1990). Mechanisms in the cycle of violence. *Science, 250,* 1678-1683.

Dunn, L. M., Dunn, L. M., & Dunn, D. M. (1997). *The Peabody Picture Vocabulary Test* (3rd edition). Circle Pines, MN: American Guidance Service, Inc.

Dunn, N. E. (1981). Children's achievement at school-entry age as a function of mothers' and fathers' teacher sets. *Elementary School Journal, 81,* 245-253.

Ebstein, R. P., Novick, O., Umansky, R., Priel, B., Osher, Y., Blaine, D., Bennett, E. R., Nemanov, L., Katz, M., & Belmaker, R. H. (1995). Dopamine D4 Receptor (D4 DR) exon III polymorphism associated with the human personality trait of novelty seeking. *Nature Genetics, 12,* 78-80.

Ge, X., Conger, R. D., Cadoret, R. J., Neiderhiser, J. M., Yates, W., Troughton, W., & Stewart, M. A. (1996). The developmental interface between nature and nurture: A mutual influence model of child antisocial behavior and parenting. *Developmental Psychology, 32,* 574-589.

Goodman, R. (1997). The strengths and difficulties questionnaire: A research note. *Journal of Child Psychology and Psychiatry, 38,* 581-586.

Gottfried, A. W. (1984). Home environment and early cognitive development: Integration meta-analysis, and conclusions. In A. W. Gottfried (Ed.), *Home environment and early cognitive development: Longitudinal research.* Orlando: Academic Press.

Grigorenko, E. L., Wood, F. B., Meyer, M. S., Hart, L. A., Speed, W. C., Shuster, A., Pauls, D. L. (1997). Susceptibility loci for distinct components of developmental dyslexia on chromosomes 6 and 15. *American Journal of Human Genetics, 60*(1), 27-39.

Hart, B., & Risley, T. R. (1995). *Meaningful differences in the everyday experience of young American children.* Baltimore, MD: Paul H. Brookes Publishing.

Hinshaw, S. P. (1992). Externalizing behavior problems and academic under-achievement in childhood and adolescence: Causal relationships and underlying mechanisms. *Psychological Bulletin, 111*(3), 387-412.

Hughes, C. H., Deater-Deckard, K., & Cutting, A. (1999). "Speak roughly to your little boy": Gender differences in the relations between parenting and preschoolers' understanding of mind. *Social Development, 8*, 143-160.

Lonigan, C. J. (1993). *Somebody read me a story: Evaluation of a shared reading program in low-income daycare.* Paper presented at the Society for Research in Child Development, New Orleans, LA.

Maccoby, E. E., & Martin, M. (1983). Socialization in the context of the family: Parent-child interaction. In E. M. Hetherington (Volume Ed.), *Handbook of Child Psychology, Vol. 4* (pp. 1-101). New York: Wiley.

Mason, J. M. (1992). Reading stories to preliterate children: A proposed connection to reading. In P. B. Gough, L. C. Ehri, & R. Treiman (Eds.), *Reading Acquisition* (pp. 215-241). Hillsdale, NJ: Erlbaum.

Moon, C., & Wells, G. (1979). The influence of home on learning to read. *Journal of Research in Reading, 2*, 53-62.

O'Connor, T. G., Deater-Deckard, K., Fulker, D. W., Rutter, M., & Plomin, R. (1998). Gene-environment correlations in late childhood and early adolescence. *Developmental Psychology, 34*, 970–981.

O'Connor, T. G., Hetherington, E. M., Reiss, D., & Plomin, R. (1995). A twin-sibling study of observed parent-adolescent interactions. *Child Development, 66*, 812-824.

Petrill, S. A., Bermont, B., & Kruming, M. (1998). *A coding scheme for parent-child language interaction.* Middletown, CT: Wesleyan University.

Pike, A., McGuire, S., Hetherington, E. M., Reiss, D., & Plomin, R. (1996). Family environment and adolescent depressive symptoms and antisocial behavior: A multivariate genetic analysis. *Developmental Psychology, 32*, 590-603.

Plomin, R. (1994). *Genetics and experience: The interplay between nature and nurture.* Thousand Oaks, CA: Sage.

Plomin, R., & Daniels, D. (1987). Why are children in the same family so different from each other? *Behavioral and Brain Sciences, 10*, 1-16.

Ramey, C. T., & Campbell, F. A. (1991). *Poverty, early childhood education and academic competence: The Abecedarian project.* Cambridge, UK: Cambridge University Press.

Reiss, D., Neiderhiser, J. M., Hetherington, E. M., & Plomin, R. (2000). *The relationship code: Deciphering genetic and social influences on adolescent development.* Cambridge, MA: Harvard University Press.

Rescorla, L., Hyson, M., Hirsh-Pasek, K., & Cone, J. (1990). Academic expectations in mothers of preschool children. *Early Education and Development, 1*, 165-184.

Robertson, C., & Salter, W. (1997). *The Phonological Awareness Test.* East Moline, IL: LinguiSystems, Inc.

Rogoff, B. (1991). *Apprenticeship in thinking: Cognitive development in social context.* Oxford, UK: Oxford University Press.

Rowe, D., & Plomin, R. (1977). Temperament in early childhood. *Journal of Personality Assessment, 41*, 150-156.

Rutter, M. (1998). Developmental catch-up, and deficit, following adoption after severe global early privation. *Journal of Child Psychology and Psychiatry, 39,* 465-476.

Scarborough, H. S. (1998). Early identification of children at risk for reading disabilities: Phonological awareness and some other promising predictors. In B. K. Shapiro, P. J. Accardo, & A. J. Capute (Eds.), *Specific feeding disability: A view of the spectrum* (pp. 75-119). Timonium, MD: York Press.

Scarborough, H. S., & Dobrich, W. (1994). On the efficacy of reading to preschoolers. *Developmental Review, 14,* 245-302.

Scarr, S. (1992). Developmental theories for the 1990's: Development and individual differences. *Child Development, 63,* 1-19.

Schaefer, E., & Edgerton, M. D. (1985). Parent and child correlates of parental modernity. In I. Sigel (Ed.), *Parental belief systems: The psychological consequences for children (1st ed.)* (pp. 287-318). Hillsdale, NJ: Erlbaum.

Silva, P. A., Williams, S., & McGee, R. (1987). A longitudinal study of children with developmental language delay at age three: Later intelligence, reading and behaviour problems. *Developmental Medicine & Child Neurology, 29,* 630-640.

Spitz, H. H. (1986). *The raising of intelligence: A selected history of attempts to raise retarded intelligence.* Hillsdale, NJ: Erlbaum.

Tessler, R., Gamache, G., & Liu, L. (1996). *The Bi-cultural Chinese-American child socialization questionnaire.* Amherst, MA: University of Massachusetts.

Thorndike, R. L., Hagen, E. P., & Sattler, J. M. (1986). *Guide for administering and scoring the fourth edition: Stanford-Binet Intelligence Scale.* Chicago, IL: Riverside.

Tizard, B. (1991). Intercountry adoption: A review of the evidence. *Journal of Child Psychology and Psychiatry, 32,* 743-756.

Tomblin, J. B., Zhang, X., Buckwalter, P., & Catts, H. (2000). The association of reading disability, behavioral disorders, and language impairment among second-grade children. *Journal of Child Psychology and Psychiatry, 41,* 473-482.

Turkheimer, E., & Waldron, M. (2000). Nonshared environment: A theoretical, methodological, and quantitative review. *Psychological Bulletin, 126,* 78-108.

Vogler, G. P., & DeFries, J. C. (1986). Multivariate path analysis of cognitive ability measures in reading-disabled and control nuclear families and twins. *Behavior Genetics, 16*(1), 89-106.

Whitehurst, G. J., Falco, F. L., Lonigan, C. J., Fischel, J. E., DeBaryshe, B. D., Valdez-Menchaca, M. C., & Caulfield, M. (1988). Accelerating language development through picture book reading. *Developmental Psychology, 24,* 552-559.

Woodcock, R. W. (1987). *Woodcock Reading Mastery Tests, Form G–Revised.* Circle Pines, MN: American Guidance Service, Inc.

Woodcock, R. W., & Johnson, M. B. (1989). *Woodcock-Johnson Tests of Achievement, Standard Battery, Form A.* Itasca, IL: Riverside Publishing Company.

Combining the Social Relations Model and Behavioral Genetics to Explore the Etiology of Familial Interactions

Beth Manke

Alison Pike

SUMMARY. This review focuses on the Social Relations Model (SRM) and a genetic extension of this model as one approach for identifying the processes by which genetic factors influence familial exchanges (Kenny & La Voie, 1984). The basic SRM and its ability to decompose dyadic measures of family interaction into actor, partner and dyadic relationship effects is described followed by findings from 3 recent studies. Results indicate that much of familial interaction is in fact relationship specific, and not due to individual-level effects. We also discuss why and how the basic SRM is enriched through the incorporation of genetically sensitive designs and present results that suggest the importance of nonshared environmental contributions for both individual and dyadic level effects. Finally, directions for future family research are proposed including the use of genetically informative designs, the collection of round-robin robin data, and the incorporation of more diverse samples. *[Article copies available for a fee from The Haworth Document Delivery Service: 1-800-HAWORTH. E-mail*

Beth Manke is affiliated with the Department of Human Development, California State University-Long Beach. Alison Pike is affiliated with the School of Cognitive and Computing Sciences, University of Sussex, Brighton, UK.

Address correspondence to Beth Manke, Department of Human Development, California State University-Long Beach, 1250 Bellflower Blvd., Long Beach, CA 90840-1602 (E-mail: bmanke@csulb.edu).

[Haworth co-indexing entry note]: "Combining the Social Relations Model and Behavioral Genetics to Explore the Etiology of Familial Interactions." Manke, Beth, and Alison Pike. Co-published simultaneously in *Marriage & Family Review* (The Haworth Press, Inc.) Vol. 33, No. 2/3, 2001, pp. 179-204; and: *Gene-Environment Processes in Social Behaviors and Relationships* (ed: Kirby Deater-Deckard, and Stephen A. Petrill The Haworth Press, Inc., 2001, pp. 179-204. Single or multiple copies of this article are available for a fee from The Haworth Document Delivery Service [1-800-HAWORTH, 9:00 a.m. - 5:00 p.m. (EST). E-mail address: getinfo@haworthpressinc.com].

179

address: <*getinfo@haworthpressinc.com*> *Website:* <*http://www.HaworthPress. com*> © *2001 by The Haworth Press, Inc. All rights reserved.]*

KEYWORDS. Social Relations Model, dyadic interactions, genetic influences

There is growing recognition that familial interactions may be due, in part, to people's general dispositions or tendencies to *behave* in a consistent manner with and *elicit* behaviors from family members. In other words, a person's behavior with family members may actually represent traits or characteristics that transcend specific relationships or contexts. These tendencies and predispositions are often referred to as child effects, actor effects, niche picking, and partner effects. Although the theoretical literature is replete with discussions of these effects, we continue to rely on single measures of familial interaction in our empirical investigations, thereby prohibiting the separation and examination of these individual level effects. For example, when we examine a sister's report of her warmth towards her brother, we are unable to determine whether the elevated warmth she reports is due to her general tendency to be a warm person, or her brother's tendency to elicit warmth. Further, we cannot discern whether the sister's warmth is not due to general dispositions of either sister or brother but represents, instead, a product of the unique relationship between siblings.

An alternative approach that allows for the examination of these various effects is the Social Relations Model (Kenny & La Voie, 1984). The Social Relations Model (SRM) involves the study of multiple two-person interactions and is based on the premise that in measures of dyadic relationships, responses from individuals about their interactions with one another are not independent of the responses obtained from other individuals reporting on the same relationship. The SRM states that each family member's interactions with other family members are a function of three independent components or sources of variance: (a) actor effects–a person's disposition or tendency to behave in a similar way toward all family members, (b) partner effects–a person's tendency to elicit similar interactions from all family members, and (c) dyadic relationship effects–factors unique to specific dyadic relationships plus measurement error.

The purpose of this review is to first describe the basic SRM as it applies to familial interactions and to delineate results from 3 recent studies. We then explain why and how it is beneficial to extend the theoretically derived phenotypic model to behavioral genetic designs in order to examine the etiology of actor, partner and dyadic relationship effects. Finally, we propose directions for future family research.

THE SOCIAL RELATIONS MODEL

As noted above, the SRM apportions variance in familial interactions into three independent sources: actor effects, partner effects and dyadic relationship effects. As an example of this approach, suppose Jennifer, an older sibling, reports that she is particularly warm to her younger brother Scott. First, this may indicate that Jennifer is generally a warm and affectionate person. That is, Jennifer may have a tendency to be warm and affectionate with *all* family members, as would be evidenced by a significant actor effect for older siblings. In other words, actor effects indicate whether individuals demonstrate cross-situational consistency in their behavior (i.e., consistency across multiple partners). Jennifer's warmth with her younger brother may also be attributed to her brother's tendency to *elicit* warmth from all family members, as would be demonstrated by a significant partner effect for younger siblings. In essence, partner effects indicate the extent to which characteristics of the partner determine the actor's behavior. A different partner, such as a mother or father, may have elicited different amounts of warmth from Jennifer.

Family interactions, such as Jennifer's warmth with her brother, may also be due to factors that are uniquely dyadic; that is, effects due to the unique relationship between two family members. The dyadic relationship effect represents, for example, the extent to which Jennifer's affection towards her brother cannot be attributed to Jennifer's general disposition toward being warm (i.e., actor effect) or her brother's tendency to elicit warmth from family members (i.e., partner effect). In essence, the relationship effect is an interaction effect as it occurs at the level of the dyad, independent of the individual-level components (i.e., actor and partner effects).

The basic SRM, which includes data from three family members (e.g., mothers, older siblings and younger siblings), is depicted as a path diagram in Figure 1. This basic model can be easily modified to incorporate additional family members such as fathers or third siblings. While the inclusion of additional family members does not alter the method for estimating actor, partner and dyadic relationship effects, it

FIGURE 1. Path diagram of Social Relations Model (SRM). OS = older sibling, YS = younger sibling, M = mother, and DR = dyadic relationship effect. The six rectangles are measured variables of directional familial interaction (e.g., OS → YS = older sibling's behavior towards the younger sibling) as reported by older siblings, younger siblings and mothers or as rated by observational coders. The actor effects are latent variables representing a person's disposition or tendency to act in a certain way to all family members. Partner effects related variables representing the tendency of family members. Partner effects are latent variables representing the tendency of family members to elicit similar interactions from all other family members. Dyadic relationship effects are latent variables representing factors unique to each particular dyadic relationship plus error of measurement. The curved, two-way arrows indicate correlations between the variables they connect, and the one-way arrows represent paths, standardized partial regressions of the SRM effect on the measured variable.

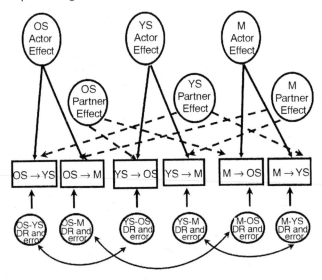

does afford greater analytic flexibility (see Kashy and Kenny, 1990 for details). In order to use the SRM, one must collect round-robin data whereby each person in the family reports on their interactions with each of the other family members. Or, in the case of observational studies, each family member must be rated in respect to their behavior toward each of the other family members.

In Figure 1, the actor, partner, and dyadic relationship effects are portrayed as circles or latent variables and are estimated by fitting the SRM to the variance-covariance matrix of the six measures of familial inter-action depicted in the rectangles (manifest variables). Note that actor and partner effects are estimated separately for *each* family member. In addition, dyadic relationship effects are calculated for each of the six re-lationships; that is, mother-older sibling, mother-younger sibling, older sibling-younger sibling, and so on. These relationship effects are direc-tional; the relationship component for the older sibling with the younger sibling may differ from the relationship component for the younger sib-ling with the older sibling. In order to separate error from dyadic rela-tionship effects, studies must incorporate replications, either over time or across measures (Kashy & Kenny, 1990). Thus, when data are col-lected at only one time point and/or include only one measurement for each dimension of familial interaction, results concerning relationship effects include measurement error and so must be interpreted with cau-tion (Ingraham & Wright, 1986).

SRM effects are estimated for the entire sample (i.e., across fami-lies), not within *each* family. A significant actor effect for older sib-lings, in our example, would indicate that older siblings differ across families in the extent to which they are consistently warm to their moth-ers, and younger siblings. That is, some older siblings consistently di-rect a great deal of warmth toward family members, whereas other older siblings consistently display little affection in familial interactions.

FINDINGS FROM THREE RECENT STUDIES

Over the past decade, the basic SRM has been used to examine vari-ous dimensions of familial interaction including power behaviors and approval statements (Cook & Dreyer, 1984), parent-child and sibling play (Stevenson, Leavitt, Thompson, & Roach, 1988), control and in-fluence (Cook, 1993), and parental affective style (Cook, Kenny, & Goldstein, 1991). Here we discuss the results from 3 additional studies, each conducted in the context of genetically informative designs and thus, as will be discussed later, were able to employ a genetic extension of the SRM. Two of the 3 studies used data drawn from the Colorado Adoption Project, a longitudinal investigation of genetic and environ-mental contributions to children's development. These investigations were based on 3-person families (i.e., mothers, older siblings and youn-ger siblings), and relied on self-report measures of familial interaction. The third study draws on data from the Nonshared Environment and Adolescent Development project, a study of identical twins, fraternal twins, full siblings, half-sibling and unrelated sibling pairs. This study incorporates both self-report and observational data, and is based on 4-person families (i.e., mothers, fathers, older siblings and younger siblings). The basic SRM results from each of the three studies are described in turn, followed by summary comments.

In the first study, familial warmth, conflict, and self-disclosure about positive and negative things were examined for the subsample of the Colorado Adoption Project (CAP: Plomin & DeFries, 1985; Stocker, Dunn, & Plomin, 1989) for which data from siblings was collected. Participants were 104 school-aged sibling pairs (44 adopted and 60 biological sibling pairs) and their mothers. Data concerning familial interactions were obtained via semistructured telephone interviews during which mothers, older siblings and younger siblings discussed how often they were warm, conflictual and self-disclosing to each of the other two family members (Manke & Plomin, 1997). Results based on the entire sample revealed that actor effects (child and mother) were significant for all four dimensions of familial interaction, suggesting that dyadic family interaction is significantly accounted for by mothers' and children's general tendency to behave similarly with family members (see Table 1). Given the reliance on self-report data, it is also possible that significant actor effects are partially attributable to a rater bias whereby family members inflate their consistency in behavior towards other family members, due perhaps to a positive or negative response tendency. In contrast to the results concerning actor effects, only the partner effects for familial conflict were significant. Finally, the results revealed pervasive significant relationship effects across all dimensions of family interaction. This suggests that family members make unique adjustments to each other that cannot be explained by actor or partner effects (Manke & Plomin, 1997).

The same sample of adoptive and nonadoptive sibling pairs, assessed 2 years later, was used in the second study (Manke, Pike, Hobson, & Plomin, 2001). In this study, familial use of humor was examined using data obtained via semistructured telephone interviews during which mothers, older siblings, and younger siblings discussed how often they used various aggressive (e.g., sarcasm, roughhousing, practical jokes) and affiliative (e.g., telling funny stories about oneself or others, laughing at embarrassing incidents, and acting silly) humor behaviors (HUMOR: Manke, 1998) with each of the other two family members. Although familial humor was attributed to both individual and dyadic level effects, results differed by type of humor. The basic SRM results in Table 2 suggest that affiliative humor is accounted for by mother and

TABLE 1. Variance Estimates from the Social Relations Model (SRM) Analyses of Familial Conflict, Warmth, and Self-Disclosure

	Conflict	Warmth	Self Disclosure About Positive Things	Self Disclosure About Negative Things
Actor Effects:				
Mother	.39*	.52*	.74*	.79*
Older Sibling	.62*	.58*	.73*	.43*
Younger Sibling	.66*	.42*	.48*	.50*
Partner Effects:				
Mother	.29*	.54*	.13	.27
Older Sibling	.42*	.25	.05	.00
Younger Sibling	.36*	.25	.03	.00
Dyadic Relationship Effects:				
M → OS	.32*	.40*	.53*	.61*
M → YS	.17	.11	.20	.53*
OS → M	.29	.55*	.83*	.99*
OS → YS	.31*	.81*	1.12*	1.32*
YS → M	.20	.27	.98*	.78*
YS → OS	.73*	.23*	1.24*	1.34*
χ^2	$\chi^2(3) = 4.61$	$\chi^2(3) = .42$	$\chi^2(3) = .23$	$\chi^2(3) = 8.58*$
GFI	.99	1.00	1.00	.97

Note. *$p < .05$. M = Mother, OS = Older Sibling, YS = Younger Sibling. For the fit of the standard model, significance is based on chi-square analysis and goodness-of-fit index (GFI). Significance of the SRM effects (actor effects, partner effects and dyadic relationship effects) is based on the chi-square change from the standard model to a reduced model in which the effect is set to zero. All chi-square values are those obtained using covariance matrices. Adapted from Manke & Plomin (1998).

child actor effects, partner effects and dyadic relationship effects, whereas aggressive humor is explained only by actor and dyadic relationship effects. This means that affiliative humor such as telling funny stories and acting silly with family members depends on the general characteristics or dispositions of both persons in the relationship, whereas the use of aggressive humor, such as the use of practical jokes and sarcasm, depends on the characteristics of the person engaging in the humorous behavior, not the partner or target of such humor.

TABLE 2. Variance Estimates from the Social Relations Model (SRM) Analyses of Humor

	Affiliative Humor	Aggressive Humor
Actor Effects:		
Mother	6.40*	6.64*
Older Sibling	16.51*	7.70*
Younger Sibling	14.96*	7.25*
Partner Effects:		
Mother	6.59*	1.54
Older Sibling	3.35^{T}	1.03
Younger Sibling	3.33*	2.86
Dyadic Relationship Effects:		
M → OS	7.47*	5.13*
M → YS	2.10	4.17*
OS → M	.00	1.21
OS → YS	8.18*	11.32*
YS → M	7.56*	3.65*
YS → OS	9.53*	7.75*
χ^2	$\chi^2(3) = 2.34$	$\chi^2(3) = 8.63*$
GFI	.99	.97

Note. $^{T}p < .10$; $*p < .05$. M = Mother, OS = Older Sibling, YS = Younger Sibling. For the fit of the standard model, significance is based on chi-square analysis and goodness-of-fit index (GFI). Significance of the SRM effects (actor effects, partner effects and dyadic relationship effects) is based on the chi-square change from the standard model to a reduced model in which the effect is set to zero. All chi-square values are those obtained using covariance matrices. Adapted from Manke, Pike, Hobson & Plomin (2001).

These results would suggest that affiliative and aggressive humor have different origins and lend support to the idea that these two types of humor serve different functions within social relationships. Affiliative humor is thought to enhance intimacy and facilitate communication and thus it is not surprising that this type of humor is accounted for by both actor and partner effects. In contrast, aggressive humor is thought to serve as an outlet for hostility, hence, it makes sense that this type of humor is accounted for largely by actor effects.

The final study extends the previous two studies in two ways (Pike, Manke, Hetherington, Reiss, & Plomin, 2001). First, fathers were included (in addition to mothers and two siblings) which allows for a more comprehensive examination of family relations. Second, this study utilized observational data (in addition to self-report data), allowing for the examination of actor effects above and beyond rating biases that may be operating when self-report measures are utilized. As mentioned previously, data for this study were drawn from the Nonshared Environment and Adolescent Development project. Participants were 702 same-sex sibling pairs aged ten to eighteen years: identical twins (n = 93), fraternal twins (n = 98), full siblings in never-divorced families (n = 95), full siblings in stepfamilies (n = 182), half siblings in stepfamilies (n = 109) and unrelated siblings in stepfamilies (n = 130). Data concerning familial aggression were obtained from videotaped observations of dyadic interactions collected during home visits. Each family member specified areas of disagreement that they then discussed during ten-minute dyadic interactions; tapes were subsequently coded using a global coding system (Hetherington & Clingempeel, 1992). In addition, all four family members reported on their own use of symbolic aggression during conflict with each of the other family members using the Conflict Tactics Scale (Straus, 1979).

Given the relatively large sample size and corresponding power to detect significant variance for the effects, it comes as no surprise that all actor, partner and dyadic relationship effects (parent and child) for both the observational and self-report measures of familial aggression/conflict were significant (see Table 3). The presence of actor effects for both observational and self-report data suggests that previously detected actor effects derived exclusively from self-report data are due to more than just rater bias. Interestingly, actor effects are about twice as large in size as compared to partner effects for both self-report and observational data. In addition, the dyadic relationship effects captured the majority of the variance in all cases.

After reviewing the results from the above three studies a few general observations can be made. From the preponderance of large dyadic relationship effects, it is clear that much of familial interaction is in fact relationship specific, and not due to the individual-level effects of actor and partner. It is also clear that actor effects predominate over partner effects, a finding that is especially persuasive given the replication across self-report and observational data.

TABLE 3. Variance Estimates from the Social Relations Model (SRM) Analyses of Observational Aggression and Self-Reported Familial Conflict

	Anger/Rejection (observational data)	Conflict (self report data)
Actor Effects:		
Mother	.18*	4.67*
Father	.19*	5.53*
Older Sibling	.23*	8.35*
Younger Sibling	.20*	11.55*
Partner Effects:		
Mother	.14*	1.09*
Father	.12*	1.15*
Older Sibling	.12*	4.57*
Younger Sibling	.09*	5.57*
Dyadic Relationship Effects:		
M → F	.56*	8.25*
M → OS	.32*	11.86*
M → YS	.38*	10.00*
F → M	.54*	10.81*
F → OS	.37*	9.56*
F → YS	.39*	7.39*
OS → M	.44*	14.36*
OS → F	.37*	14.06*
OS → YS	.51*	24.55*
YS → M	.43*	12.05*
YS → F	.42*	14.07*
YS → OS	.55*	22.84*
χ^2	$\chi^2 (48) = 197.57*$	$\chi^2 (48) = 292.14*$
RMSEA	.07	.09

Note. *$p < .05$. M = Mother, F = Father, OS = Older Sibling, YS = Younger Sibling. For the fit of the standard model, significance is based on chi-square analysis and the Root Mean Square Error of Approximation (RMSEA). Significance of the SRM effects (actor effects, partner effects and dyadic relationship effects) is based on the standard errors. All chi-square values are those obtained using covariance matrices. Adapted from Pike, Manke, Hetherington, Reiss, & Plomin, 2001.

Genetic Extension of the SRM

Although the basic SRM contributes crucially to our understanding of interactions among family members because it apportions variance in familial interaction data into actor, partner, and dyadic relationship effects, it is impossible, within the context of the basic SRM, to examine the *etiology* of these effects. That is, we cannot determine the processes that drive individual and dyadic level effects. A first step toward understanding these processes would be to employ a genetic extension of the SRM.

Although genetic and environmental influences on SRM effects have been investigated in only 3 studies to date, considerable effort in the past twenty years has been devoted to estimating genetic contributions to more traditional measures and models of familial interaction. The conclusion that emerges from the numerous studies employing diverse methods (e.g., adoption and twin designs) and measures (e.g., questionnaires and videotaped observations) is that parent-child and sibling interactions show significant and often substantial genetic influence (e.g., Dunn & Plomin, 1986; McGuire, Manke, Eftekhari, & Dunn, 2000; O'Connor, Hetherington, Reiss, & Plomin, 1995; Plomin, Reiss, Hetherington, & Howe, 1994; Rende, Slomkowski, Stocker, Fulker, & Plomin, 1992; Rowe, 1981).

At a conceptual level, genetic influences on familial interactions are typically interpreted as due to relatively *stable* genetically influenced traits that are expressed across situations and relationships. More specifically, genetic influences are thought to accrue because family members' behavior is reflected in, or affected by, genetically influenced partner and actor effects. That is, children may *elicit* interactions from family members because of genetically influenced partner effects such as temperament, psychopathology or cognitive ability. For example, an easygoing child who displays few conduct problems may elicit warmth from family members. Children may also actively *create* interactions with family members based on genetically influenced actor effects. For example, a child who is highly emotional may actively seek out conflictual exchanges with other family members. Although less frequently discussed, it has also been proposed that genetic influences on familial interactions are *not* due to general dispositions such as actor and partner effects, but are instead unique to experience itself, that is, experiences such as the emotionally rich and intense interactions in the context of the family (Plomin, 1995).

Despite the fact that our interpretations of genetic influence on familial interactions allude to the presence of actor, partner and (less often) dyadic relationship effects, traditional behavioral genetic analyses that examine single measures of familial interaction cannot test these conclusions empirically. That is, traditional univariate genetic analyses do not allow for the simultaneous investigation of multiple family relationships, or the examination of genetic and environmental influences on trait-like characteristics such as actor and partner effects *separately* from genetic and environmental contributions to dyad-specific interactions.

The ability to examine genetic and environmental influences on actor, partner and dyadic relationship effects *separately* is the unique value of applying a genetic extension to the basic SRM. For the first time, we are able to clarify previous results concerning genetic influences on familial interaction. Significant genetic influences on actor effects, for example, would suggest that previously detected genetic contributions are due, at least in part, to genetically influenced trait-like characteristics that transcend specific relationship situations.

Genetically informative samples are necessary in order to use the genetic extension of the SRM. That is, sibling pairs or parent-child pairs of varying genetic relatedness must be included. Suitable samples might include identical and fraternal twins, adopted and biological full sibling pairs, full siblings and unrelated siblings, or adoptive and biological parents and their children. In a child-based design (where the sibling pairs vary in genetic relatedness), genetic and environmental contributions to the SRM effects can be determined for the children's actor and partner effects, and the dyadic relationship effects between siblings. SRM effects involving the mother, such as the mother actor and partner effects, are not examined in the context of this genetic model. It should be noted that the inclusion of data concerning parent-child interactions in the SRM is, nevertheless, necessary for the estimation of child effects that are used in the genetic analyses. When parents and their offspring who vary in genetic relatedness are included in a study (as is the case in studies of adoptive and biological parents and their children), genetic influences are assessed for both parent *and* child effects, as well as the dyadic relationship effects between parents and children. Because child-based designs are more common, and because all previous applications of the SRM and the genetic extension have utilized child-based designs, we limit our discussion of the genetic extension to this design.

As a first step in estimating genetic contributions, the correlations between siblings of varying genetic relatedness are examined and compared. Because one cannot correlate unreliable variance, it is necessary that all variance estimates examined in the genetic analyses be reliably greater than zero. Thus, variance estimates for both older and younger sibling effects (for all sibling types) must be significant in order to run the genetic extension. For example, in order to estimate genetic and environmental influences on actor effects for warmth in the sibling adoption design, the actor effects for older siblings in nonadoptive families, older siblings in adoptive families, younger siblings in nonadoptive families and younger siblings in adoptive families must *all* be significant, otherwise genetic analyses are invalid.

Sibling correlations are estimated by re-running the basic SRM analyses for the different sibling types separately, so that the correlations between the sibling SRM effects of interest can be calculated. Genetic influence is implied if the following pattern emerges among the correlations: identical twins > fraternal twins and full siblings > half siblings (i.e., children who have the same biological mother *or* biological father, but not both) > unrelated siblings (e.g., step siblings, adopted siblings). The genetic effects size, heritability, can be estimated from several of these comparisons, such as doubling the difference between the correlations for MZ twins (who are identical genetically), and DZ twins (whose genetic relatedness is 50% on average) or quadrupling the difference between the correlations for full siblings (.50 genetic relatedness) and half siblings (.25 genetic relatedness).

Although our focus thus far has been on estimating genetic influences, environmental contributions can also be assessed by examining the sibling correlations. Shared environmental influences index those environmental factors that work to make siblings similar and are most clearly observed in the correlation between unrelated siblings who can be similar for no other reason than environmental factors. Shared environmental factors are also implied if the correlations among the different sibling pairs are significant and similar. Finally, nonshared environmental influences can be estimated. Nonshared influences refer to the remainder of the variance not explained by genetics or shared environment; this variance is attributed to environmental influences unique to each sibling. If nonshared environmental influences are substantial, we would expect the correlations for all sibling pairs to be low.

Rather than estimating heritability and environmental influences in a piecemeal manner by comparing sibling correlations, model fitting is a more elegant and comprehensive approach. The basic univariate behavioral genetic model is described elsewhere in more detail (see Plomin & DeFries, 1985). An example of this model for assessing genetic and environmental contributions to sibling actor effects is depicted as a path diagram in the upper portion of Figure 2. Analogous genetic models have also been developed for the investigation of sibling partner and sibling dyadic relationship effects. In order to conduct the genetic analyses, both the basic SRM (lower portion of Figure 2) and the genetic extension (upper portion of Figure 2) are run as one model, using secondary latent variables to estimate the genetic and environmental components.

The model in Figure 2 illustrates that resemblance between siblings (e.g., similarity in actor effects) can be due to two independent latent factors: effects of genes (G) and shared environmental influences (E_s).

FIGURE 2. Path diagram of genetic model designed to examine genetic and environmental contributions to sibling actor effects. OS = older sibling, YS = younger sibling. G is the latent variable representing genetic variance. E_s is the latent variable representing shared environments influence common to a sibling pair. E_n represents residual variance that does not result in sibling covariance. The curved, two-way arrows indicate correlations between the variables they connect, and the one-way arrows represent paths, standardized partial regressions of the Social Relations Model (SRM) effect on the latent variable. The lower unmarked portion of this figure represents the basic SRM and is described in Figure 1.

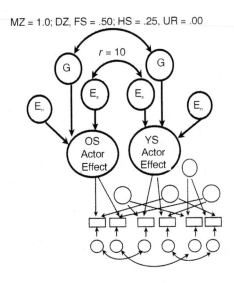

The genetic effects for siblings are correlated to the extent that siblings are genetically similar. The shared environmental effects for siblings are correlated 1.0 for all sibling pairs reared in the same home because, by definition, siblings (genetically related and unrelated) share the same family environment. The residual arrow (E_n) for each sibling refers to nonshared environmental influences unique to each individual. These environmental influences are not connected by a two-headed arrow because, again by definition, these effects are specific to an individual and operate to make siblings different, not similar. Interestingly, traditional genetic investigations of single measures of familial interaction are unable to estimate *true* nonshared environmental influences separate from measurement error because these two sources of variance are confounded. As a result, the legitimacy of nonshared environmental influences has been questioned. In contrast to traditional genetic analyses, application of a genetic model to the SRM allows us to assess true nonshared environmental influences on actor and partner effects because these effects are latent variables free from measurement error. Actor and partner effects are estimated with the relationship effects partialled out, and it is here, in the relationship effects, where residual error variance is contained.

Genetic Extension Results

Although the basic SRM has been used to examine several dimensions of familial interaction, only 3 studies have been conducted in the context of genetically informative designs and thus were able to employ a genetic extension of the SRM. We reviewed the basic SRM results of these 3 studies in a previous section; what follows are the results from the genetic extensions.

In this first study to employ a genetic extension of the SRM (Manke & Plomin, 1997), genetic analyses could only be conducted for actor effects associated with conflict and self-disclosure about positive things, and dyadic relationship effects associated with warmth and self-disclosure about positive and negative things. The criterion for inclusion in the genetic analyses was not met by any of the other effects across the four dimensions of familial interaction. That is, at least one of the effects (actor, partner or dyadic relationship effect) either in the nonadoptive or adoptive families was nonsignificant.

The sibling correlations by adoptive status and the model fitting results are presented in Table 4. The sibling correlations for actor effects for both conflict and self-disclosure about positive things suggest ge-

TABLE 4. Sibling Correlations by Adoptive Status and Maximum-Likelihood Model Fitting Results for Familial Conflict, Warmth and Self-Disclosure

	Sibling Correlations		Variance Components				
	NA (n = 60)	A (n = 44)	h^2	es^2	en^2	χ^2 (df)	GFI
Conflict:							
Actor Effects	.61*	.47*	.41^T	.41*	.18	39.54* (26)	.88
Warmth:							
Dyadic Relationship Effects	.27^T	.24	.03	.43*	.54*	53.80* (27)	.92
Self Disclosure About Positive Things:							
Actor Effects	.37^T	.14	.20^T	.16	.64*	27.41 (26)	.93
Dyadic Relationship Effects	.25	.19	.02	.26^T	.72*	28.17 (27)	.92
Self Disclosure About Negative Things:							
Dyadic Relationship Effects	.40*	.41*	.06	.38*	.56*	28.38 (27)	.91

Note. ^{T} p < .10, *p < .05. NA = nonadoptive sibling pairs, A = adoptive sibling pairs. For the fit of the standard model, significance is based on chi-square. Significance for the heritability estimate (h^2) is based on the chi-square change from the standard model to a reduced model in which G is set to zero. Significance for the shared environmental estimate (es^2) is based on chi-square change from the standard model to a reduced model in which E_s is set to zero. Significance for the nonshared environmental estimate (en^2) is based on a chi-square change from the standard model to a reduced model in which E_n is set to zero. GFI = goodness of fit index. All chi-square values are those obtained using covariance matrices. Adapted from Manke & Plomin (1998).

netic influence; the nonadoptive sibling correlations are significant in both cases and exceed the sibling correlations for adoptive pairs. Although shared environmental influences are implicated for actor effects associated with conflict (as evidenced by a significant correlation for adoptive sibling pairs), nonshared environmental influences appear chiefly responsible for actor effects associated with self-disclosure, as indicated by generally modest sibling correlations for both adoptive and nonadoptive pairs. The sibling correlations for dyadic relationship effects associated with warmth and self-disclosure about positive and negative things suggest some dyadic reciprocity in the unique adjustments that siblings make to each other, and that these adjustments are due to environmental influences, both shared and nonshared. Genetic influences appear minimal in that nonadoptive sibling correlations are not substantially greater than those for adoptive sibling pairs.

For the most part, maximum likelihood model-fitting analyses confirm the impressions gleaned from the correlational results. That is, genetic contributions appear significant, albeit at the level of a trend, for actor effects, but not dyadic relationship effects. Shared environmental influences play an important role in dyadic relationship effects (for all dimensions) and for actor effects associated with conflict. For most of the SRM effects, nonshared environmental influences are also significant and substantial, explaining over half of the variance in most cases. The estimate of significant nonshared environmental influence for actor effects associated with self-disclosure about positive things is especially interesting as actor effects are free from measurement error. Because the nonshared environmental contributions to dyadic relationship effects are still confounded with measurement error, nonshared environmental influences on these effects must be interpreted with caution.

In the second study to employ a genetic extension of the SRM (Manke, Pike, Hobson, & Plomin, 2001), actor and dyadic relationship effects for both affiliative and aggressive humor met the criteria for inclusion in the genetic analyses. Sibling correlations for these SRM effects by adoptive status are shown in Table 5. For affiliative humor actor and dyadic relationship effects, the moderate yet similar correlations for both nonadoptive and adoptive siblings suggest the presence of shared environmental influences. Genetic influences are not implicated as the nonadoptive sibling correlations *do not* exceed the correlations for adoptive sibling pairs. In contrast, for aggressive humor, genetic influences are implicated for both sibling actor and dyadic relationship effects as evidenced by larger nonadoptive sibling correlations than adoptive sibling correlations. Substantial environmental influences of the nonshared type are implicated for sibling actor and dyadic relationship effects for both affiliative and aggressive humor as indicated by generally moderate correlations for both nonadoptive and adoptive sibling pairs.

The maximum likelihood model-fitting results support the interpretations drawn from the correlational findings. That is, genetic influence was found for aggressive humor whereas shared environmental effects were significant for affiliative humor. Nonshared environmental contributions were significant and substantial for both types of familial humor.

Although the models for affiliative humor demonstrate a good fit (as evidenced by nonsignificant chi-squares), chi-squares for the aggres-

TABLE 5. Sibling Correlation by Adoptive Status and Maximum-Likelihood Model Fitting Results for Familial Humor

	Sibling Correlations		Variance Components				
	NA (n = 56)	A (n = 40)	h^2	es^2	en^2	χ^2 (df)	GFI
Affiliative Humor:							
Actor Effects	.26*	.34*	.09	.29*	.62*	33.41 (26)	.90
Dyadic Relationship Effects	.29^T	.20^T	.00	.38*	.62*	38.87 (27)	.88
Aggressive Humor:							
Actor Effects	.20	−.37*	.36*	.00	.64*	41.57* (26)	.91
Dyadic Relationship Effects	.34*	.19	.43^T	.16	.41*	42.41* (27)	.90

Note. * p < .05, ^{T}p < .10. NA = nonadoptive sibling pairs; A = adoptive sibling pairs. For the fit of the standard model, significance is based on chi-square. Significance for the heritability estimate (h^2) is based on the chi-square change from the standard model to a reduced model in which G is set to zero. Significance for the shared environment estimate (es^2) is based on chi-square change from the standard model to a reduced model in which E_s is set to zero. Significance for the nonshared environmental estimate (en^2) is based on chi-square change from the standard model to a reduced model in which E_n is set to zero. GFI = goodness-of-fit index. All chi-square values are those obtained using covariance matrices. Adapted from Manke, Pike, Hobson, & Plomin (2001).

sive humor models are significant, suggesting that the results for aggressive humor should be interpreted with caution. Further, it should be noted that the significant heritability estimate for actor effects for aggressive humor is due, in part, to the negative sibling correlation for adoptive sibling pairs. This means that the adoptive siblings' behavior is not merely independent of each other (as one would expect of two randomly chosen adolescents), but that an adolescent's behavior can actually be predicted as opposing that of his or her sibling. What might be at work is a process known as sibling contrast or deidentification effects for adoptive siblings whose similarity is not boosted by genetic relatedness, as is the case for nonadoptive siblings (Carey, 1986). That is, in the absence of genetic similarity, adoptive siblings may purposely carve out separate roles in the family, or directly compete with one another in the family context for scarce resources (e.g., parental attention). These siblings may work to engage in different levels of aggressive humor, thereby resulting in the negative sibling correlation for this type of humor use. In short, siblings may represent reference points against which they can judge, and if necessary change, their behavior and may contribute in important ways to sibling differences in humor use. For nonadoptive siblings, such a contrast may be overridden by genetic similarity in humor use. It is interesting to note that sibling contrast effects have been found for other adolescent characteristics such as temperament (Saudino, McGuire, Reiss, Hetherington, & Plomin, 1995).

In the final study to employ a genetic extension to the basic SRM, both observational and self-report data on familial anger/rejection and control were examined (Pike, Manke, Hetherington, Reiss, & Plomin, 2001). The sibling correlations by sibling type are presented in Table 6; maximum likelihood model fitting results are still in progress. The predominately cascading sibling correlations (i.e., MZ > DZ = FI = FS > HF > UN) for actor effects associated with observational anger/rejection and partner effects for self-reported conflict indicate the presence of moderate to substantial genetic influence. Genetic contributions are negligible or modest for the other effects, as the magnitude of the sibling correlations do not consistently increase with the degree of genetic relatedness. Shared environmental influences are implicated for anger/rejection partner effects (observational data), actor effects for conflict (self-report data), and for dyadic relationship effects for both types of data. That is, for these effects, the sibling correlations are significant and similar in magnitude across sibling types and suggest that growing up in the same family, regardless of genetic heritage, makes siblings similar. Nonshared environmental influences appear important in varying degrees for actor, partner and dyadic relationship effects for both observational and self-report data as suggested by sibling correlations less than unity. It should be noted, however, that the substantial MZ-twin correlations for many of the effects indicates that these estimates of nonshared environmental influence will be smaller than those reported in the first two studies.

Given the genetic results from the 3 studies above, we can draw a few general conclusions concerning the etiology of individual and dyadic level effects. With one exception (i.e., aggressive humor), dyadic relationship effects appear to be due to environmental influences, both shared and nonshared, whereas genetic contributions are nonsignificant. This means that the unique adjustments siblings make to one another that cannot be explained by either child's actor or partner effects are due to environmental factors. Shared environmental influences appear less important for actor effects as significant shared environmental contributions were detected for only two dimensions of familial interaction:

TABLE 6. Sibling Correlations by Sibling Type for Observational Aggression and Self-Reported Familial Conflict

	MZ (n = 93)	DZ (n = 98)	FI (n = 95)	FS (n = 182)	HF (n = 109)	UN (n = 103)
Anger/Rejection (observational data):						
Actor Effects	.57*	.40*	.30*	.45*	.30*	.15*
Partner Effects	.01	.34*	.55*	.48*	.23*	.43*
Dyadic Relationship Effects	.63*	.72*	.45*	.67*	.55*	.56*
Conflict (self-report data):						
Actor Effects	.46*	.42*	.08	.22*	.16*	.26*
Partner Effects	.82*	.72*	.61*	.46*	.51*	.28*
Dyadic Relationship Effects	.64*	.21*	.33*	.33*	.16*	.42*

Note. *p < .05. MZ = identical twins, DZ = fraternal twins, FI = full siblings in two-parent never divorced families, FS = full siblings in step-families, HF = half siblings in step-families, UN = unrelated siblings in step-families. Adapted from Pike, Manke, Hetherington, Reiss, & Plomin (2000).

self-reported conflict in both the CAP and NEAD samples and affiliative humor. In contrast, significant genetic contributions to actor effects are implicated for several measures of familial interaction (both self-report and observational) including aggressive humor, conflict, anger/rejection and self-disclosure about positive things. Because genetic and environmental contributions to partner effects were examined in only one study, it is hard to draw any conclusions concerning these effects.

In summary, we can see that the basic SRM and the genetic extension contribute crucially to our understanding of familial interactions and shed light on some of the common assumptions made by behavioral geneticists when interpreting traditional behavioral genetic analyses. First, much of familial interaction is in fact relationship specific, and not due to individual-level effects, as is often assumed when behavioral genetic studies of unitary measures of familial interaction are interpreted. Our findings are in line with a recent theory suggesting that the context of close relationships is the salient process through which predispositions are translated into effective experiences and developmental outcomes (Reiss, Neiderhiser, Hetherington, & Plomin, 2000).

Further, these dyadic relationship effects, along with actor effects, are due primarily to nonshared environmental effects. This means that the environmental contributions of greatest importance are those that work to make siblings different, rather than similar. Although nonshared environmental estimates for dyadic relationship effects contain measurement error, actor and partner effects are free from this type of error. This is an important contribution to traditional behavioral genetic research as the identification of true nonshared environmental influences free from measurement error provides legitimacy to the study of systematic nonshared environmental factors (Turkheimer & Waldron, 2000). But what can account for these large nonshared environmental contributions? Likely candidates surely include children's experiences outside the family such as interactions with peers, teachers and romantic partners.

Results from the basic SRM further suggest that actor effects predominate over partner effects. This is of interest when typical explanations for genetic influence on parental behavior are considered. Such influences are often interpreted as being due to the consistent elicitation of particular qualities of behavior by children (partner effects). The phenotypic findings from the basic SRM, in combination with the genetic results, however, reveal that consistent elicitation is not an important source of variability in family interactions.

Although the three studies described above provide valuable information concerning genetic and environmental contributions to individual and dyadic level effects, they do not provide simple answers. We must look to future investigations to clarify which dimensions of familial interaction are accounted for by actor, partner and dyadic relationship effects, and the degree to which these effects are due to environmental and genetic contributions. In the final section of this paper, we outline several directions for future work.

FUTURE DIRECTIONS

In this paper we have argued that the SRM and its genetic extension help us answer the question, "What accounts for familial interaction?" We recognize, however, that the SRM will be useful only to the degree that other researchers understand its utility, have data suitable for its application, and feel comfortable with its use. At a minimum, this means that it will be important for family researchers and behavioral geneticists to adopt round-robin designs with which detailed (and directional) information is gathered concerning *all* dyadic family relationships. The reliance on single measures of familial interaction is no longer acceptable if we are to understand what drives family interactions. In short, multiple measurements of family relationships across multiple sources are necessary if we are to disentangle variance attributable to actor, partner and dyadic relationship effects.

In practical terms, this means that researchers must make the decision to use the SRM *before* the initiation of a study, as most researchers do not collect round-robin data unless there is a specific reason to do so. For many this will require a sacrifice of breadth for depth in data collection, and will represent a difficult decision especially when there is limited time with each participant and/or observational data are of interest. In four-person families (e.g., mothers, fathers, older siblings and younger siblings) there are twelve dyadic combinations, each of which must be observed and then coded twice in order to obtain directional information concerning all dyadic familial interactions.

In addition to collecting round-robin data, it will be important to include siblings of varying degrees of genetic relatedness, as it is impossible to disentangle environmental and genetic contributions to actor, partner, and dyadic relationship effects without multiple sibling types (e.g., identical twins, fraternal twins, full siblings, etc.). Including siblings of differing genetic heritage is easier than once thought given the naturally occurring family types in today's society. For example, in a current sibling study in a large metropolitan area, families were recruited where the criterion for inclusion stipulated that all participating families must have at least two children between the ages of 8 and 14, and that the children be living with a parent willing to participate (Manke, Robertson, O'Brien, MacDonald, Wyche, & Berglund, 1997). Single, two-parent and stepfamilies were allowed to participate, although no specific family or sibling type was targeted. The resulting sample included twins, full siblings, half siblings and unrelated siblings. Given a large enough sample, the inclusion of these multiple sibling types would allow for the separation of genetic and environmental contributions to individual and dyadic level effects. Further, the utilization of multiple sibling types might prove especially useful in situations where specific SRM effects for one or two sibling types do not meet the criteria for inclusion in the genetic analyses (i.e., variances are nonsignificant). In these situations, one or two sibling types could be omitted without rendering the genetic analyses invalid.

Although not mandatory, future applications of the SRM should also incorporate short-term longitudinal data using the same measure, or use multiple measures of each construct at one time point. The inclusion of such data will allow for the estimation of dyadic relationship effects separate from measurement error. Further, the incorporation of multiple assessments of the same construct would allow researchers to clarify the extent to which nonshared environmental influences on dyadic relationship effects are the product of real substantive influences as opposed to measurement error.

Although it is difficult to imagine that dyadic relationship effects are merely the reflection of measurement error, we can draw no firm conclusions regarding such effects as long as they are confounded with error.

Yet another design consideration concerns the nature of our samples. That is, we need to incorporate more diverse samples in our investigations of individual and dyadic level effects if we are to evaluate the generalizability of previous SRM findings and understand the extent to which changes in genetic and environmental influences on familial relationships are a function of differing contexts. While the call to include more diverse samples is made in almost every discussion of empirical results derived from relatively advantaged White middle-class two-parent samples, paying lip service to its importance is no longer sufficient. Increased sample diversity can be obtained in a variety of ways. First, one could examine a larger set of relationships including fathers, extended kin, and fictive kin. This may be particularly important in contexts where children live with, or have frequent contact with, people who provide crucial social support such as grandparents, aunts, uncles and cousins. It would be interesting to investigate whether the variance in familial interaction attributable to actor, partner and dyadic relationship effects is similar when different types of family relationships are considered. For example, in families where children live with their mothers and grandparents (as compared to families where children live with two parents and no extended kin), perhaps a greater portion of the variance in mother-child interactions is attributable to the unique adjustments parent and child make to each other.

The second way in which sample diversity may be obtained is through the inclusion of more ethnically and socioeconomically diverse families. For example, work is currently underway to examine individual and dyadic level effects associated with familial conflict in a sample of African American, Latino and Non-Hispanic White families. Given that African American parents place a greater importance on children's deference to authority figures (Manke, Robertson, Halgunseth, Chau & Wozniak, 2000), it will be interesting to see whether the same pattern of actor, partner and dyadic relationship effects for familial conflict emerge across ethnic groups. Other diverse samples might include extreme environmental contexts such as domestically violent families or parents and children living in poverty.

Finally, it will be important for family researchers to recognize that the SRM and its genetic extension are only the *first* steps in pinpointing processes responsible for differences in family relationships. The next step is to move beyond mere partitioning of variance to understanding *specific* family processes. Of course the SRM and the genetic extension provide a road map or compass for investigating specific processes, suggesting which factors hold the most promise for further investigation. For example, the finding that children's actor effects for aggressive humor may not be influenced by shared environmental factors advises against further examination of environmental factors common to siblings. Instead, future research would do better to examine factors that are known to be genetically influenced or environmental factors experienced differently by siblings. Although previous attempts to link genetic contributions to personality with single measures of familial interaction have been met with limited success, it is possible that genetically influenced traits may be more consistently related to actor and partner effects which are isolated from the variance attributed to measurement error and unique dyadic relationship effects.

REFERENCES

Carey, G. (1986). Sibling imitation and contrast effects. *Behavior Genetics, 16,* 319-341.

Cook, W. L. (1993). Interdependence and the interpersonal sense of control: An analysis of family relationships. *Journal of Personality and Social Psychology, 64,* 587-601.

Cook, W. L., & Dreyer, A. (1984). The social relations model: A new approach to the analysis of family-dyadic interaction. *Journal of Marriage and the Family, 46,* 679-687.

Cook, W. L., Kenny, D. A., & Goldstein, M. J. (1991). Parental affective style risk and the family system: A social relations model analysis. *Journal of Abnormal Psychology, 100,* 492-501.

Dunn, J., & Plomin, R. (1986). Determinants of maternal behavior toward three-year-old siblings. *British Journal of Developmental Psychology, 4,* 127-137.

Hetherington, E. M., & Clingempeel, W. G. (1992). Coping with marital transitions: A family systems perspective. *Monographs of the Society for Research in Child Development, 57,* (2-3, Serial No. 227).

Ingraham, L. J., & Wright, T. L. (1986). A cautionary note on the interpretation of relationship effects in the social relations model. *Social Psychology Quarterly, 449,* 93-97.

Kashy, D. A., & Kenny, D. A. (1990). Analysis of family research designs: A model of interdependence. *Communication Research, 17,* 462-482.

Kenny, D. A., & La Voie, L. (1984). The social relations model. In L. Berkowitz (Ed.), *Advances in experimental social psychology* (pp. 141-182). Orlando, FL: Academic Press.

Manke, B. (1998). Genetic and environmental contributions to children's interpersonal humor. In W. Ruch (Ed.), *The sense of humor: Explorations of a personality characteristic* (pp. 361-384). New York: Mouton de Gruyter.

Manke, B., Pike, A., Hobson, R., & Plomin, R. (2001). *Humor in adolescent family relations: A genetic study of individual and dyadic relationship effects.* Manuscript submitted for publication.

Manke, B., & Plomin, R. (1997). Adolescent familial interactions: A genetic extension of the social relations model. *Journal of Social and Personal Relationships, 14,* 505-522.

Manke, B., Robertson, R., O'Brien, K., MacDonald, P., Wyche, C., & Berglund, J. (1997). *Developing multi-ethnic family studies on a shoestring: Emphasizing university-community collaborations.* Paper presented at the Ethnicity and Development Conference: Integrating Cross-Cultural and Developmental Perspectives, Lincoln, NE, November.

Manke, B., Robertson, R., Halgunseth, L., Chau, Y., & Wozniak, R. (2000). *Understanding family values: The role of ethnicity and social class.* Paper presented at the biennial meeting of the Society for Research on Adolescence, Chicago, IL, April.

McGuire, S., Manke, B., Eftekhari, A., & Dunn, J. (2000). Children's perceptions of sibling conflict during middle childhood: Issues and sibling (dis)similarity. *Social Development, 9,* 173-190.

O'Connor, T. G., Hetherington, E. M., Reiss, D., & Plomin, R. (1995). A twin-sibling study of observed parent-adolescent interactions. *Child Development, 66,* 812-829.

Pike, A., Manke, B., Hetherington, E. M., Reiss, D., & Plomin, R. (2001). *Families with adolescents: The social relations model applied within a genetically sensitive design.* Manuscript in preparation.

Plomin, R. (1995). Genetics and children's experiences in the family. *Journal of Child Psychology and Psychiatry, 36,* 33-68.

Plomin, R., & DeFries, J. C. (1985) *Origins of individual differences in infancy.* New York: Academic Press.

Plomin, R., Reiss, D., Hetherington, E. M., & Howe, G. (1994). Nature and nurture: Genetic influences on measures of family environment. *Developmental Psychology, 30,* 32-43.

Reiss, D., Neiderhiser, J. M., Hetherington, E. M., & Plomin, R. (2000). *The relationship code: Deciphering genetic and social influences on adolescent development.* Cambridge, MA: Harvard University Press.

Rende, R. D., Slomkowski, C.L., Stocker, C., Fulker, D.W., & Plomin, R. (1992). Genetic and environmental influences on maternal and sibling interaction in middle childhood: A sibling adoption study. *Developmental Psychology, 28,* 484-490.

Rowe, D. C. (1981). Environmental and genetic influences on dimensions of perceived parenting: A twin study. *Developmental Psychology, 17,* 203-208.

Saudino, K. J., McGuire, S., Reiss, D., Hetherington, E. M., & Plomin, R. (1995). Parent ratings of EAS temperaments in twins, full siblings, half siblings, and step siblings. *Journal of Personality and Social Psychology, 68,* 723-733.

Stevenson, M. B., Leavitt, L. A., Thompson, R. H., & Roach, M .A. (1988). A social relations model analysis of parent and child play. *Developmental Psychology, 24,* 101-108.

Stocker, C., Dunn, J., & Plomin, R. (1989). Sibling relationships: Links with child temperament, maternal behavior, and family structure. *Child Development, 60,* 715-727.

Straus, M. A. (1979). Measuring intrafamily conflict and violence: The Conflict Tactics (CT) Scales. *Journal of Marriage and the Family, 41,* 75-85.

Turkheimer, E., & Waldron, M. (2000). Nonshared environment: A theoretical, methodological, and quantitative review. *Psychological Bulletin, 126,* 78-108.

The Friends of Siblings:
A Test of Social Homogamy
vs. Peer Selection and Influence

Michael S. Gilson
Cathleen B. Hunt
David C. Rowe

SUMMARY. Using the full siblings and their friends in the *National Longitudinal Study of Adolescent Health* (Add Health), we examined two possible processes of friends' similarity: selection/influence and social homogamy. For two phenotypes, delinquency and verbal intelligence (VIQ), self-reports of siblings and their friends were linked by a computer algorithm. Structural equation models were then applied to a 4 × 4 covariance matrix (siblings 1 and 2, friends of sibling 1, friends of sibling 2) for each phenotype separately. The best-fit model of delinquency included only a selection/influence process and no social homogamy process. In

Michael S. Gilson, Cathleen B. Hunt, and David C. Rowe are affiliated with the University of Arizona.

Address correspondence to: David C. Rowe, School of Family and Consumer Sciences, Campus Box 210033, University of Arizona, Tucson, AZ 85721 (E-mail: dcr091@ag.arizona.edu).

The National Longitudinal Study of Adolescent Health (Add Health) was designed by J. Richard Udry and Peter Bearman and funded by Grant P01-HD31921 from the National Institute of Child Health and Human Development. Data sets can be obtained by contacting the Carolina Population Center, 123 West Franklin Street, Chapel Hill, NC 27516-3997 (E-mail: jo_jones@unc.edu).

[Haworth co-indexing entry note]: "The Friends of Siblings: A Test of Social Homogamy vs. Peer Selection and Influence." Gilson, Michael S., Cathleen B. Hunt, and David C. Rowe. Co-published simultaneously in *Marriage & Family Review* (The Haworth Press, Inc.) Vol. 33, No. 2/3, 2001, pp. 205-223; and: *Gene-Environment Processes in Social Behaviors and Relationships* (ed: Kirby Deater-Deckard, and Stephen A. Petrill) The Haworth Press, Inc., 2001, pp. 205-223. Single or multiple copies of this article are available for a fee from The Haworth Document Delivery Service [1-800-HAWORTH, 9:00 a.m. - 5:00 p.m. (EST). E-mail address: getinfo@haworthpressinc.com].

contrast, the best-fit model of VIQ supported a social homogamy process. Friends' resemblance in delinquency rates can be attributed to selection and/or influence. However, friends' intellectual resemblance may be due to friends' placements in ability tracks in schools and in other social contexts instead of being a result of a preference for a particular VIQ level. *[Article copies available for a fee from The Haworth Document Delivery Service: 1-800-HAWORTH. E-mail address: <getinfo@haworthpressinc. com> Website: <http://www.HaworthPress.com> © 2001 by The Haworth Press, Inc. All rights reserved.]*

KEYWORDS. Peer selection, social homogamy, siblings, IQ, delinquency

During adolescence friends are similar in race, age, gender, marijuana use, political orientation, educational aspirations, delinquency, and in other characteristics (Kandel, 1978). What is true of friends is also true of spouses. As with friends, spouses are similar in their demographic characteristics and matched on their personality traits, educational levels, IQs, and attitudes. They also assort positively for delinquency (Vandenberg, 1972; Krueger, Moffitt, Caspi, Bleske et al., 1998). An important theoretical question is, what are the causes of behavioral resemblance between spouses and friends? Although this article will focus on the resemblance of friends' phenotypes, it is useful to contrast possible pathways to resemblance for both friends and spouses. A large research literature already exists on spouses' behavioral resemblance, and many of the issues raised in this literature also can apply to our theoretical understanding of friends' behavioral resemblance.

PROCESSES OF BEHAVIORAL RESEMBLANCE

In this section, we use the term "partners" to refer either to a pair of friends or to a husband/wife pair. One source of possible similarity for partners is a *selection process*. A selection process occurs when the characteristics, or phenotype, of one person is attractive to another person so that this individual is chosen as a partner. This process implies both that a particular phenotype is noticed and discriminated and that it leads to an active preference for another individual. Selection is probably a source of partners' resemblance on invariant demographic characteristics. For instance, race and age are characteristics that exist prior to a pairing of two individuals, and are not altered by it. Partners tend to be similar in age and racial group membership. On phenotypes such as IQ, which are relatively stable traits over many years, behavioral resemblance is also possibly due to selection. In a social psychology study (Newcomb, 1961), college students' similarity in social attitudes was experimentally manipulated by assigning them to room with someone similar or dissimilar to themselves in their social attitudes. Roommates holding similar attitudes became friends; dissimilar ones did not, which demonstrates a process of selection with experimental control.

A second source of behavioral resemblance is an *influence process*. An influence process occurs when individuals who are initially dissimilar in their characteristics are paired and then become more alike because one person accepts the influence of the other. In studies of married couples, influence can be inferred when spouses who were initially dissimilar become more alike over the years of their marriage. In friendship pairs, influence may be captured by a longitudinal design that examines whether initially dissimilar friends become more alike with a longer duration of friendship.

A third source of behavioral resemblance is a *social homogamy* process. A key feature of a social homogamy process is that partners' behavioral resemblance is incidental upon their general social circumstances. That is, people do not actively judge one another's phenotypes and then, on the basis of their preferences, reject some individuals as partners while accepting others. Nor do they change in response to another's influence. Rather, the social circumstances in which people exist would lead to behavioral resemblance even if the individuals within these circumstances were choosing partners entirely at random. A social homogamy process is important in behavioral genetic studies because it and a selection process have different consequences for modeling kinship correlations; specifically, a selection process can induce complex correlations between additive genetic effects and shared environmental effects across generations (for details, see Neale & Cardon, 1992).

For any study of partners' behavioral resemblance to be valid, data on phenotypes must be obtained directly from each individual in a pair. Thus, we can exclude from consideration any study that used individuals who reported on their own phenotypes as well as those of their partners. This restriction therefore excludes the great majority of studies purporting to inform the scientific community about partners' resemblance, because most studies have just one informant for each pair.

The Behavioral Resemblance of Spouses

The majority of studies of spousal assortment (Vandenberg, 1972; Price & Vandenberg, 1980; Yamaguchi & Kandel, 1993) have found a selection process to be stronger than an influence process for most phenotypes. Two observations that compellingly support this inference are: (1) newlywed couples and dating couples were similar for many phenotypes, and (2) their degree of similarity was not moderated by years of cohabitation or marriage.

Behavioral genetic (BG) studies are able to distinguish a selection from a social homogamy process using data on the phenotypes of twin offspring and their parents (Eaves, Fulker, & Heath, 1989; Neale & Cardon, 1992). Neale and Cardon applied both phenotypic selection and social homogamy models to a scale of fear of social criticism (the spouse correlation was .21). A selection process fit well, whereas a social homogamy process gave unrealistic parameter estimates because shared environmental effects on the fear scale were too weak to make a social homogamy model realistic; a social homogamy model necessarily implies the existence of a shared environmental influence on a phenotypes' variation.

Using adult twins and their spouses, other BG models can distinguish a selection from a social homogamy process. Phenotypes on full-siblings and their spouses also can be used to make this same distinction. In a sample of Hawaiians, Nagoshi and Johnson (1994) applied BG structural equation models to data on personality and attitudinal phenotypes from full siblings and their spouses. An active phenotypic selection process led to spousal similarity on the Eysenck Personality Questionnaire Lie scale, Education, and Language Use. A social homogamy process was found for Radicalness-Conservatism and pidgin language use. Tambs, Sundet, and Berg (1993) applied a BG model to 138 pairs of identical twins and their spouses, with the phenotype being IQ. The best-fit model indicated that IQ resemblance resulted from a social homogamy process rather than from a selection process. Nagoshi, Johnson, and Ahern (1987) found that spouses' resemblance on education was determined by both processes, whereas their similarity on verbal ability was mostly due to a social homogamy process. As this short review demonstrates, the relevant processes are likely to vary among phenotypes, but a social homogamy process cannot be neglected when considering the origins of spouses' similarity. This observation carries an important implication for studies of friendship, almost none of which have considered a social homogamy process as an alternative explanation to an influence or a selection process.

Friends' Behavioral Resemblance

The friendship literature provides ample support for both influence and selection processes, which are not mutually exclusive explanations of friends' behavioral resemblance. In the classic study of this topic, Kandel (1978) examined 957 friendships over a school year. Friendships formed at the beginning of the year already possessed some similarity in characteristics such as marijuana use, political orientation, educational aspirations, and minor delinquency. Students who remained friends for the full year also become more alike in their phenotypes. On the basis of her analyses, Kandel concluded that about half of friends' behavioral resemblance was due to influence and half was due to selection. Ennett and Bauman (1994) and Fisher and Bauman (1988) used social network analyses, combined with longitudinal designs, to study cigarette smoking and alcohol consumption in groups of friends. In these studies, adolescents identified their best friends at two points in time, and information on substance use was linked among participants and friends. Support for an influence process was found as initially nonsmokers or nondrinkers paired with smoker or drinker friends were almost twice as likely to initiate use over the time period. However, smokers and drinkers also were more likely to acquire smoker and drinker friends than nonsmokers and nondrinkers, indicating a selection process. In contrast to Kandel's study, Bauman and his colleagues concluded that a selection process was stronger than an influence process on friends' resemblance for smoking and alcohol consumption. In past studies of these behaviors, the use of a single informant who may inaccurately report on friends' characteristics has greatly overstated the strength of an influence process (Bauman & Ennett, 1994).

In this article, we apply a model of a social homogamy process vs. influence/selection process to two phenotypes—delinquency and verbal IQ—using data from full siblings and their friends. We combine the influence and selection processes because we cannot include reciprocal paths in a structural equation model that uses data from full siblings only; if other kinships were added (e.g., identical twins), a fuller specification of the model would be possible. An innovative aspect of this study is to consider a social homogamy process as a source of friends' phenotypic similarity, a process that has already been shown to be important for spouses' similarity but remains largely unexplored in adolescent friendships.

The two phenotypes were chosen as possibly contrasting. The data reviewed above on spouses indicated that social homogamy created their similarity for verbal IQ and some related variables. Although IQ has not been examined specifically, friends are similar on related phenotypes. For example, Hamm (2000) found that adolescent friends from an ethnically diverse sample shared similar academic orientations. Adolescents also chose friends who were highly similar to themselves in academic performance (Tesser, Campbell, & Smith, 1984) and who were similar in their choice of academic subjects (Virk, 1983). Because IQ is a highly stable lifetime trait, friends' intellectual similarity could be a result of an active selection process—people wanting to be with others equally smart as themselves. Friends could also influence one another's intellectual abilities. Certainly, this is the assumption behind peer tutoring efforts favored by many schools (Gyanani & Pahuja, 1995). Yet one also can see a possible effect of social homogamy because most junior high and high schools track students by ability groups. Thus, even if students were to pick friends randomly from their classmates, some matching would occur due to this preexisting ability stratification created by school policies.

Nearly every study of delinquency to examine the issue has found that friends are similar in their rates of delinquency and in the rates of other mildly deviant behaviors (e.g., underage smoking and drinking). Those studies with a longitudinal design have demonstrated that both selection and influence contribute substantially to friends' similarity, with their relative influence varying among studies and specific phenotypes. No prior study of which we are aware has examined whether a social homogamy process contributes to friends' behavioral resemblance on delinquency.

METHOD

Sample and Procedure

The National Longitudinal Study of Adolescent Health (Add Health) was designed to assess the health status of adolescents and explore the causes of adolescent health-related behaviors. Add Health began with a total sample of over 90,000 adolescents surveyed in school. The primary sampling frame was all high schools in the United States that had an 11th grade and had an enrollment of at least 30 students. A random sample of eighty high schools was selected from this sampling frame, taking into consideration enrollment size, region, school type, ethnicity, and urbanicity. The largest feeder school for each high school was also included in the sample. Seventy-nine percent of the schools initially contacted agreed to participate. Schools that refused to participate were replaced by another school in the same sampling stratum, resulting in a final sample of 134 schools. Within these schools, 90,118 of 119,233 eligible students (75.6%) in grades 7 to 12 completed a self-administered instrument for optical scanning.

Using both school roster information and information provided by adolescents during the school interview, a random sample of 15,243 adolescents was also selected for a detailed home interview. This sample was stratified by gender and age. The in-home interview was completed by 12,188 (79.5%) of these adolescents. In addition to this core sample, a number of subsamples were also selected for the home interview. These subsamples include samples of disabled adolescents, adolescents from well-educated African American families, adolescents from typically understudied racial and ethnic groups, and a special sibling pairs sample. Overall, 20,745 adolescents completed the in-home interview. Details on the Add Health Study have been reported elsewhere (Udry & Bearman, 1998).

Included in the 20,745 adolescents is the Add Health *pairs sample* (N = 3,139 sibling pairs) that was selected using information from the in-school questionnaires and school rosters. A probability sample of full siblings was drawn (N = 1,251). The average age in both the full sample and the sibling pairs sample is approximately 16 years (SD = 1.7). Although the pairs sample also contains identical twins, fraternal twins, half-siblings, and unrelated siblings, these samples were not used because of the large amounts of missing data due to merging of friend and sibling data.

The Add Health study connected youths' identifiers to create social networks of friends, romantic partners, and peers. Network pairs come from 16 schools in which all enrolled students were selected for the in-home interviews. These were two large schools (with a total combined enrollment of over 3,300) and 14 small schools (with enrollments fewer than 300). One of the large schools is predominately white and located in a mid-sized town. The other is ethnically heterogeneous and is located in a major metropolitan area. The 14 small schools have various characteristics. They are located in rural and urban areas. Some are public schools and some are private. All of the respondents in the Add

Health "saturated school" sample analyzed here were given the opportunity to nominate up to ten friends and were given rosters with the names of all students in their school to aid in selection. Three hundred respondents from these schools, however, were erroneously not given the opportunity to nominate up to ten friends, but instead were asked to nominate one best friend of each sex.

The measure used here was mean score of the first nominated male friend and first nominated female friend on delinquency and verbal IQ. Both sexes were used to increase the likelihood that both male and female respondents would have an identified friend in the sample. Pairwise deletion of cases with missing data was used to maximize sample size. The friends phenotypes analyzed here qualify for the modeling effort because they are not proxies for phenotypes collected through the self-report of the respondent but are instead collected directly through the self-report of the friend.

The data reported here come from the in-home Wave I questionnaire. This in-home interview was completed between April and December 1995. All respondents were given the same interview, which took from one to two hours to complete. All data were recorded on laptop computers. Sensitive questions were asked via audio files drawn off the hard disk to coincide with presented questions. In addition to maintaining data security, the audio-casio procedure minimized the potential for interviewer bias.

Measures

Delinquency. This scale assessed the frequency of total delinquency during the past year, using 15 items. Responses ranged from 0 = never to 1 = "1 or 2 times" to 2 = "3 or 4 times" to 3 = "5 or more times." The items included minor to serious delinquent behaviors typical of self-report instruments, such as "Take something from a store without paying for it," "Get into a serious physical fight," and "Steal something worth more than $50." Item scores were averaged, with higher scores representing a higher rate of delinquency. The range was 0 to 3. The Cronbach's coefficient alpha was .85. About 25% of siblings had delinquency scores of zero. Because the friend variable was an average of two individuals' scores, only about 10% of scores on friends' delinquency were zero. The delinquency distribution was *J-shaped*, with a positive skewness, as is typical of self-reported delinquency scales. The delinquency scores were square root transformed to reduce this skewness. For each sibling, the effect of gender was statistically removed by using residuals from the regression of delinquency on gender.

Add Health Peabody Picture Vocabulary Test (Verbal IQ). This test is an abridged version of the Peabody Picture Vocabulary Test-Revised. It was administered at the beginning of the in-home interview. This test of vocabulary involved the interviewer reading a word aloud. The respondent then selected an illustration that best fits the word. Each word had four, simple, black-and-white illustrations arranged in a multiple-choice format from which the respondent indicated his or her choice. For example, the word "furry" had illustrations of a parrot, a dolphin, a frog, and a cat from which to choose. There were 78 items on the verbal IQ test, and raw scores have been standardized by age. Individuals with VIQ scores below 50 were dropped from the data set.

Whites had a higher VIQ mean than either African Americans or Hispanics. This racial difference would tend to bias sibling correlations because siblings usually share the same racial self-identification. Because siblings and friends were also nearly always of the same race, the racial differences also would increase these correlations. For this reason, the VIQ scores were corrected for their association with racial group by using residuals from the regression of VIQ on two dummy variables, one representing whites vs. African Americans and the other representing whites vs. Hispanics. Thus, race-corrected VIQ was used to calculate the covariance matrices. Although these results are not shown, similar findings emerge when the uncorrected scores are used.

The structural equation model used here to distinguish these various theories of friends' similarity is illustrated by Figure 1. Four observed scores are in the model. They are the phenotype of sibling 1, the phenotype of sibling 2, that of two friends of sibling 1, and that of two friends of sibling 2. The model requires two latent variables. One is labeled F for *family* and represents the total influence on the family that makes siblings alike, a combination of shared environmental and genetic influences. With more groups, the genetic and shared environmental contributions to the variation of a phenotype could be separated, but our analysis included only biological full-siblings. The path coefficient e represents the familial effect. The latent variable SH stands for social homogamy. It is uncorrelated with familial influences by assumption. Equal paths d from it to siblings' and friends' phenotypes account for its contribution to phenotype variation. The third path in the figure, c, connects a sibling with a friend. The path c includes both influence and prior phenotypic selection; they are inseparable in this model. Error terms on the phenotype of sibling 1 and 2 were constrained to be equal;

FIGURE 1. Structural Equation Model for Phenotypic Resemblance (P SIB = Sibling's Phenotype, P FND = Friend's Phenotype, SH = Social Homogamy, and F = Family Influences)

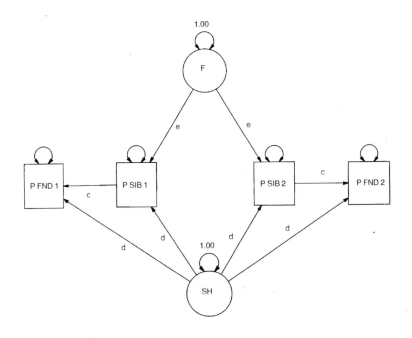

an equality constraint was also placed on the friends' error terms. The full model can be simplified by omitting any one path, *e*, *d*, or *c*, to test with one degree of freedom (df) whether that path is necessary to the model. A chi square greater than 3.84 would indicate a loss of fit when a path coefficient is omitted. No specific predictions were made.

RESULTS

The results are organized into two sections. The first presents the findings on delinquency; the second, those on the verbal intelligence. When the different models are introduced, it should be noted that the models used for both sets of variables were structured identically. The full model (Figure 1) incorporated influence/selection as well as social homogamy. Removing path *c* from the full model yielded a social homogamy model. Removing path *d* from the full model created a selection/influence model.

Delinquency

Table 1 provides the correlations between each of the four delinquency variables below the diagonal. Statistically significant correlations were found between sibling 1 and sibling 1's friends' delinquency ($r = .23$, $p < .001$) as well as between sibling 2 and sibling 2's friends' delinquency ($r = .19$, $p < .01$). Also significant was the correlation between sibling 1's and sibling 2's delinquency scores ($r = .19$, $p < .05$). No significant relationships were found between delinquency scores of friends 1 and friends 2, or between a sibling and the other sibling's friends. The near zero correlation between friends 1 and friends 2 eliminates a social homogamy process. The .19-.23 correlations between siblings and their friends fits with an expectation of the selection/influence model, that *c* exceed 0. The sibling correlation of .19 indicates that *e* is also greater than 0.

Table 1 also presents the 4 × 4 covariance matrix used in fitting our structural equation model. As pair-wise deletion generated these covariances, the sample sizes varied. The sample size used in the structural equation models was determined by averaging the *N*s for four vari-

TABLE 1 . Covariances, Variances, and Correlations Between Siblings' and Friends' Delinquency

Variable	1	2	3	4
1. Del SIB 1	.103	.021	.019	−.001
	407			
2. Del FND 1	.23***	.084	.003	.005
	258	325		
3. Del SIB 2	.19*	.03	.099	.016
	169	135	403	
4. DEL FND 2	−.02	.06	.19**	.071
	133	136	255	327

Note. Covariances are presented above diagonal, variances on the main diagonal (bold), and correlations below diagonal. FND = friend; SIB = sibling. <u>N</u>s are in italics.
* $p < .05$
** $p < .01$
*** $p < .001$

ances and six covariances (mean $N = 254$). The covariance matrix was fit to the full and reduced structural equation models in Figure 1 using the Mx computer program (Neale, 1997).

Table 2 gives the fit indices from the structural equation models. The full model gave an excellent fit according to its nonsignificant chi square ($p = .60$). The submodels were evaluated by a one df change in chi square test. Dropping d resulted in a trivial chi square change ($\Delta \chi^2 = .04$). Thus, the social homogamy process was unnecessary for the model to fit well. On the other hand, dropping the selection/influence process parameter c greatly worsened model-fit ($\Delta \chi^2 = 12.48$, df $= 1$, p $< .01$). The two other fit statistics confirmed these results. A small and negative Akaike (1974) information criterion statistic indicates a good fit. The Root Mean Squared Error Approximation, or RMSEA (Neale, 1997), is a goodness-of-fit statistic that is relatively independent of sample size. It is a function of the chi square, sample size, and degrees of freedom. According to Neale (1997), values under .10 indicate a good fit and values below .05 an excellent fit. All models had good fits by this criterion, but those of the full and the selection/influence process model (.00) were nearly perfect and better than that of the social homogamy process model (.08).

The standardized path coefficients for the best-fitting selection/influence process model are shown in Figure 2. Thus, the expected correlation of sibling–> friend delinquency is .20 under the model, the value of the standardized path coefficient. The expected delinquency correlation of friend 1 and friend 2, following Wright's rules for tracing path diagrams (Loehlin, 1987), is $.20^2 \times (.44)^2 = .008$. Lastly, the expected correlation of full siblings' delinquency is $(.44)^2 = .19$. These values agree closely with the observed correlations in Table 1.

TABLE 2. Fit Indices of Path Model for Delinquency

Model	df	χ^2	p	AIC	RMSE
Full	5	3.68	.60	−6.32	.00
Social Homog. c = 0	6	16.16	.01	4.16	.08
Select./Influen. d = 0	6	3.72	.71	−8.27	.00

Note. Mean $N = 254$, mean sample size over Ns from four variances and six covariances, AIC = Akaike's information criterion (Akaike, 1974). RMSEA = Root Mean Square Error Approximation.

FIGURE 2. Selection/Influence Structural Equation Model for Delinquency

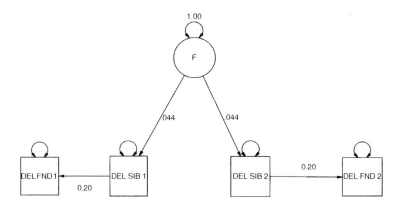

Verbal Intelligence

The covariances and correlations between each of the four, race cor-rected VIQ variables are shown in Table 3. Most variables were signifi-cantly correlated with one another *(p < .001)*. The sibling correlation was *r* = .45. Correlations between friends were smaller, ranging from .25 to .36. In contrast to delinquency, a substantial association existed between the VIQs of friend 1 and friend 2 *(r* = .32). Hence, the correlational structure was consistent with a social homogamy process making friends and siblings alike.

Mx was used to fit the structural equation model to the covariance ma-trix presented in Table 3 (mean *n* = 241). The covariance matrix was again fit to the full, social homogamy, and selection/influence process models and was assessed by the same fit indices as before. According to the fit statistics in Table 4, both the full model and the social homogamy process model yielded excellent fits. The selection/influence process model was rejected, however. It yielded both a statistically significant χ^2 and also an increase in χ^2 in comparison to the full model ($\Delta \chi^2 = 22.1$, $df = 1$, $p < .001$). The increase in χ^2 between the full model and the social homogamy process model was only 1.49 ($df = 1$). Thus, we accepted this model as the best-fit, most parsimonious description of the covariance matrix.

Figure 3 presents the standardized path coefficients from the best fit-ting social homogamy process model. This model yields an estimated

TABLE 3. Covariances, Variances, and Correlations Between Siblings' and Friends' VIQ

Variable	1	2	3	4
1. VIQ SIB1	**146.7**	32.5	66.4	45.0
	389			
2. VIQ FND 1	.25*	**119.9**	22.4	41.1
	243	*317*		
3. VIQ SIB 2	.45*	.17	**146.0**	50.7
	154	*127*	*383*	*237*
4. VIQ FND 2	.32*	.32*	.36*	**138.9**
	124	*126*	*234*	*314*

Note. Covariances are presented above diagonal, variances on the main diagonal (bold), and correlations below diagonal. FND = friend; SIB = sibling. *N*s are in italics.
Race statistically controlled.
* *p* < .001

TABLE 4. Fit Indices of Path Models for Verbal IQ

Model	df	χ^2	p	AIC	RMSEA
Full	5	6.41	.27	−3.6	.03
Social Homog.	6	7.90	.25	−4.1	.04
Select./Influen.	6	30.0	.00	16.0	.12

Note. Mean *N* = 241, mean sample size over *N*s from four variances and six covariances. AIC = Akaike's information criterion (Akaike, 1974). RMSEA = Root Mean Square Error Approximation.

sibling correlation on verbal intelligence of $.55^2 + .44^2 = .50$. Under this model, the expected VIQ correlation between friend 1 and friend 2 is $.55 \times .51 = .28$. For clarification, these standardized path coefficients *d* were unequal because the equality constraint was imposed only on the unstandardized path coefficients; the unequal error variances for siblings vs. friends enter into their calculation. The estimated VIQ correlations corresponded closely with the observed VIQ correlations presented in Table 3.

FIGURE 3. Social Homogamy Structural Equation Model for Verbal Intelligence (VIQ)

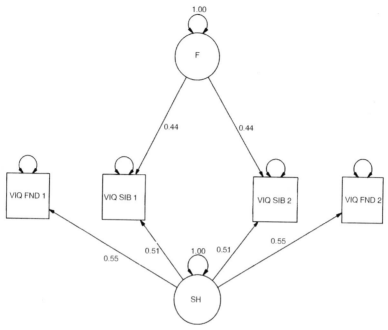

DISCUSSION

Friends are similar in a variety of their demographic characteristics, including age and race, but also in more complex phenotypes such as drug use, delinquency, and IQ. We were interested in the processes leading to friends' behavioral resemblance. In this study, two specific phenotypes were investigated: delinquency and verbal IQ. We were able to specifically examine two process models–selection/influence and social homogamy–using data from siblings and their friends. A social homogamy process refers to a similarity induced because friends share similar social niches; no active preference for the friends' phenotype is required to make the friends alike phenotypically. A selection process refers to similarity induced when friends prefer one another's characteristics; the choice of a friend is an active one. An influence process occurs if friends interact with one another so that their characteristics may change. For each sibling in the *National Longitudinal Study of Adolescent Health*, a first chosen-female friend and first chosen-male friend were identified through a linking algorithm. The friends' scores were averaged to increase sample size; thus, unlike many studies of friends using just one informant, our behavioral data come from each individual in a friendship independently. Friends (a sibling and the two named friends) were more alike in verbal IQ, with a correlation of .25 to .36, than in delinquency, with $r = .19$ to .23.

With four variables (sib1, friends1, sib2, friends2), we could model a social homogamy process using structural equations, but we could not make a distinction between selection and influence. Although we tested our hypotheses with structural equation models, intuitively a difference between a selection/influence process and a social homogamy process can be captured by a correlation of siblings' friends. If social homogamy is at work, then these friends, although they are most often different people, can be much alike in behavior because they all share one social niche. If a selection/influence process is at work, a younger sibling may resemble his friends, but the friends of an older brother or sister, who were chosen independently, may be very different in behavior from the younger sibling's friends.

A social homogamy process provided the best fitting statistical model for verbal intelligence. One explanation for a strong social homogamy process on VIQ is that friends were drawn from classmates and from individuals in school clubs and activities. American junior and senior high schools typically track students by ability level. Thus, classmates are already somewhat matched for VIQ, and even if students were to ignore intelligence when they make a friendship choice, friends will be still alike in VIQ if they are in the same ability track. Similarly, some voluntary school activities–such as chess or science clubs–would produce prior selection on VIQ, making friends alike. An interesting question is whether VIQ would become a variable for active friendship choice, if schools were to stop tracking students by ability levels. This idea might be evaluated by looking at friendship choices within a nontraditional school that avoided ability-tracking students.

Our result that favors social homogamy also accords well with the findings on assortative mating on IQ as reviewed above. In several studies, spouses' resemblance on IQ and related traits was determined by social homogamy. In early marriage, ability tracking in secondary schools may account for spouses' resemblance in IQ. However, because the age of first marriage is often in the mid-twenties, it is likely that college attendance vs. nonattendance, and attendance in more academically selective colleges, may impose the social context that leads to positive marital assortment on IQ (Kalmijn, 1998).

In contrast to verbal intelligence, our results on delinquency favored a selection/influence process, because the observed correlations on delinquency gave an excellent statistical fit to this model. For example, friends correlated about .20 on delinquency, but the delinquency of friend1 and friend2 did not correlate at all. A social homogamy process on delinquency could be statistically rejected. We view this as indicating both selection and influence: selection because childhood behavioral disorders can predict adolescent delinquency long before adolescent friendship groups are formed (Moffitt, 1993); influence because gangs and peer groups can actively encourage the antisocial behavior of their members.

One limitation of our study is that we cannot separate a selection from an influence process. In light of the studies reviewed above, selection is an important source of friends' similarity, equal to or exceeding the contribution of friends' mutual influence. Our model could be extended to separate selection and influence in two ways. One would be through the addition of longitudinal data on siblings. Even with longitudinal data, though, causality is disputable because an assumption is made that an initial score is an adequate control for all prior influences leading to friendship, which may be untrue. Another approach to estimating these effects is to add genetic kinship groups, such as half-siblings and twins. With multiple groups, bidirectional paths could be placed between friends and siblings. As a social influence cannot cause siblings' degree of genetic relatedness, the latent genetic variable can serve as an instrumental variable for a process of selection. The reverse arrow, from the friends to the sibling, would represent the effect of influence. A multiple group analysis could apportion the phenotypic variance of delinquency and verbal intelligence onto genetic, shared environmental, and nonshared environmental determinants.

In summary, our results have several important implications. One is that the use of a single informant may give misleading estimates of friends' resemblance. In studies of delinquency, friends' delinquency may correlate as high as .40 to .60 when one individual is the informant on friends' or peers' delinquency. Our estimated correlation, .20, is less than half that typically reported from single informant data (and in percentage of variance, less than 1/4th the value). This effect size must be further reduced, however, because some of the ~ .20 correlation is due to selection. On verbal intelligence, friends were less alike than siblings, with a correlation ~ .30. The idea that IQ is not particularly determinative of friendship choice seems to run counter to common sense; people tend to aggregate with others of similar IQ, in ability tracks in high school and in neighborhoods balkanized by social class in adulthood. Our data suggests that this assortment is not a result of an active choice process, at least for adolescent friendships. We suspect that our result may be wrong in a certain sense. No preference may be exhibited for IQ because adolescents may find at hand a pool of intellectually-matched peers available for friendships. Experimental or longitudinal data are needed to demonstrate a preference for similar IQ. It might also be interesting to look the IQ variability within sets of close friends of target individuals. Perhaps a lesson of spouse assortment studies, that we accommodate to others more than we really change them, may apply to friendships as well.

REFERENCES

Akaike, H. (1974). A new look at the statistical model identification. *IEEE Transactions on Automatic Control, 19*, 716-723.

Bauman, K. E., & Ennett, S. T. (1994). Peer influence on adolescent drug use. *American Psychologist, 49*, 820-822.

Eaves, L. J., Fulker, D. W., & Heath, A. C. (1989). The effects of social homogamy and cultural inheritance on the covariances of twins and their parents: A LISREL model. *Behavior Genetics, 194*, 113-122.

Ennett, S. T., & Bauman, K. E. (1994). The contribution of influence and selection to adolescent peer group homogeneity: The case of adolescent cigarette smoking. *Journal of Personality & Social Psychology, 67*, 653-663.

Fisher, L. A., & Bauman, K. E. (1988). Influence and selection in the friend-adolescent relationship: Findings from studies of adolescent smoking and drinking. *Journal of Applied Social Psychology, 18*, 289-314.

Gyanani, T. C., & Pahuja, P. (1995). Effects of peer tutoring on abilities and achievement. *Contemporary Educational Psychology, 20*, 469-475.

Hamm, J. V. (2000). Do birds of a feather flock together? The variable bases for African American, Asian American, and European American adolescents' selection of similar friends. *Developmental Psychology, 36*, 209-219.

Kalmijn, M. (1998). Intermarriage and homogamy: Causes, patterns, trends. *Annual Review of Sociology, 24*, 395-421.

Kandel, D. B. (1978). Homophily, selection, and socialization in adolescent friendships. *American Journal of Sociology, 84*, 427-436.

Krueger, R. F., Moffitt, T. E., Caspi, A., Bleske, A., & Silva, P. A. (1998). Assortative mating for antisocial behavior: Developmental and methodological implications. *Behavior Genetics, 28*, 173-186.

Loehlin, J. C. (1987). *Latent variable models: An introduction to factor, path, and structural analysis.* Hillsdale, NJ: Lawrence Erlbaum.

Moffitt, T. E. (1993). Adolescence-limited and life-course-persistent antisocial behavior: A developmental taxonomy. *Psychological Review, 100*, 674-701.

Nagoshi, C. T., & Johnson, R. C. (1994). Phenotypic assortment versus social homogamy for personality, education, attitudes, and language use. *Personality and Individual Differences, 17,* 755-761.

Nagoshi, C. T., Johnson, R. C., & Ahern, F. M. (1987). Phenotypic assortative mating vs. social homogamy among Japanese and Chinese parents in the Hawaii family Study of Cognition. *Behavior Genetics, 17,* 477-485.

Neale, M. C. (1997). *Mx: Statistical modeling.* Box 126 MCV, Richmond, VA 23298, Department of Psychiatry, 4th Edition.

Neale, M. C., & Cardon, L. R. (1992). *Methodology for genetic studies of twins and families.* Boston: Kluwer Academic Press.

Newcomb, T. M. (1961). *The acquaintance process.* New York: Holt, Rinehard & Winston.

Price, R. A., & Vandenberg, S. G. (1980). Spouse similarity in American and Swedish couples. *Behavior Genetics, 10,* 59-71.

Tambs, K., Sundet, J. M., & Berg, K. (1993). Correlations between identical twins and their spouses suggest social homogamy for intelligence in Norway. *Personality and Individual Differences, 14,* 279-281.

Tesser, A., Campbell, J., & Smith, M. (1984). Friendship choice and performance: Self-evaluation maintenance in children. *Journal of Personality & Social Psychology, 46,* 561-574.

Udry, J. R., & Bearman, P. S. (1998). New methods for new research on adolescent sexual behavior. In R. Jessor (Ed), *New perspectives on adolescent risk behavior* (pp. 241-269). New York: Cambridge University Press.

Vandenberg, S. G. (1972). Assortative mating, or who marries whom? *Behavior Genetics, 2,* 127-157.

Virk, J. (1983). Mutual attraction and friendship formation: II. *Indian Psychologist, 2,* 44-49.

Yamaguchi, K., & Kandel, D. (1993). Marital homophily on illicit drug use among young adults: Assortative mating or marital influence? *Social Forces, 72,* 505-528.

Observation of Externalizing Behavior During a Twin-Friend Discussion Task

Leslie D. Leve

SUMMARY. In the last decade, numerous collaborations between developmentalists and behavior geneticists have extended developmental theories to genetically informative samples. This study integrates the approaches used by behavior geneticists and environmental researchers to further our understanding of the etiology of children's externalizing behavior. First, genetic and environmental influences on externalizing behavior are reviewed and studies using data from parent reports, self-reports, teacher reports, public records, and observation are described. Second, observational data from a middle-childhood twin sample are presented. Results suggest significant dominant genetic effects on observed twin negative behavior during a friend interaction ($N = 148$ twin pairs). In addition, nonshared environmental effects were substantial; when controlling for the coparticipant's behavior, nonshared environ-

Leslie D. Leve is affiliated with the Oregon Social Learning Center, 160 East 4th Avenue, Eugene, OR 97401-2426 (E-mail: lesliel@oslc.org).

The author wishes to thank Allen Winebarger, John Reid, and Hill Goldsmith for their contributions to the Oregon Twin Project, and Matthew Rabel for his editorial assistance. The author also wishes to thank all of the twin and friend families who generously volunteered to participate in this research.

This research was supported by Grant No. RO1 MH54248 and Grant No. P50 MH46690, NIMH, U.S. PHS to John B. Reid; by Grant No. RO1 MH51560, NIMH, U.S. PHS to H. Hill Goldsmith; by Grant No. RO3 MH57053 and Grant No. RO1 MH 37911, NIMH, U.S. PHS to Leslie D. Leve; by Grant No. T32 MH18935, NIMH, U.S. PHS to Allen A. Winebarger; and by research funds from the University of Oregon Department of Psychology.

[Haworth co-indexing entry note]: "Observation of Externalizing Behavior During a Twin-Friend Discussion Task." Leve, Leslie D. Co-published simultaneously in *Marriage & Family Review* (The Haworth Press, Inc.) Vol. 33, No. 2/3, 2001, pp. 225-250; and: *Gene-Environment Processes in Social Behaviors and Relationships* (ed: Kirby Deater-Deckard, and Stephen A. Petrill) The Haworth Press, Inc., 2001, pp. 225-250. Single or multiple copies of this article are available for a fee from The Haworth Document Delivery Service [1-800-HAWORTH, 9:00 a.m. - 5:00 p.m. (EST). E-mail address: getinfo@haworthpressinc.com].

ment was the sole source of variation. Context and method effects are discussed. *[Article copies available for a fee from The Haworth Document Delivery Service: 1-800-HAWORTH. E-mail address: <getinfo@haworthpressinc.com> Website: <http://www.HaworthPress.com> © 2001 by The Haworth Press, Inc. All rights reserved.]*

KEYWORDS. Genetics, peers, externalizing, observational methods

Literally thousands of research articles describe the onset and developmental course of externalizing behaviors in young children (e.g., Brook, Whiteman, Finch, & Cohen, 1996; Deater-Deckard, Dodge, Bates, & Pettit, 1998; Fagot & Leve, 1998; Fergusson, Horwood, & Lynskey, 1994; Keenan & Shaw, 1995; Loeber & Dishion, 1983; McMahon, 1994; Patterson, 1986; Patterson, Capaldi, & Bank, 1991; Plomin, Nitz, & Rowe, 1990; Stoolmiller, Patterson, & Snyder, 1997; Vitaro, Brendgen, Pagani, Tremblay & McDuff, 1999). The coercion model (Patterson, 1982; Patterson, Reid, & Dishion, 1992) is one of the most widely tested theoretical models of child and adolescent externalizing behavior.[1] Coercion theory states that the primary pathway leading to child and adolescent externalizing behavior begins when a child is reinforced for responding aversively to terminate undesired parental and sibling behaviors (Patterson, 1976). The more uncooperative a child becomes, the less likely that child is to receive attention and positive feedback from the parent when appropriate behaviors are exhibited. A child receiving abundant negative reinforcement for aversive behaviors and little positive reinforcement for appropriate behaviors will likely encounter major difficulties in academic and peer settings when entering middle childhood and early adolescence (Patterson et al., 1992). In turn, such difficulties can lead to a cascading set of problems for the child, for the family, and for the community (Dishion, McCord, & Poulin, 1999).

In the last decade, numerous collaborative efforts between developmentalists and behavior geneticists have extended developmental theories such as the coercion model to genetically informative samples (e.g., Deater-Deckard, 2000; Ge et al., 1996; Kim, Hetherington, & Reiss, 1999; Leve, Winebarger, Fagot, Reid & Goldsmith et al., 1998; Neiderhiser, Reiss, Hetherington, & Plomin, 1999; Pike, McGuire, Hetherington, Reiss & Plomin, 1996; Reiss, Neiderhiser, Hetherington, & Plomin, 2000; Rutter, 1997). By including siblings, twins, and/or adoptive families, researchers can examine the contributions of genetic and environmental factors in the development of externalizing problems within a specified theoretical model. This integrative approach capitalizes on the strengths of both fields. However, methodology has been a major challenge to integrating the two fields. Often, environmental researchers rely on observational approaches to study microsocial interaction processes, and behavior genetic researchers rely on questionnaire methodology to attain the sample sizes needed to detect environmental and Gene × Environment (G × E) interaction effects (e.g., Neale & Cardon, 1992). Thus, collaborative studies that examine externalizing behavior with a genetically sensitive design *and* observational data are few and far between.

In a recent meta-analysis of 24 genetically informative studies, Miles and Carey (1997) reported just 2 studies that used observational methodology to assess aggressive behavior: Plomin, Foch, and Rowe (1981) and Rende, Slomkowski, Stocker, Fulker and Plomin (1992). Genetic studies including both observational and questionnaire data are even more sparse, showing modest to moderate correlations between observational and questionnaire data on externalizing behavior (Leve et al., 1998). These methodological issues have led developmentalists and behavior geneticists to urge that more attention be paid to the measurement of aggression (e.g., Maccoby, 2000; Plomin et al., 1990; Plomin, Reiss, Hetherington, & Howe, 1994).

In this study, I have aimed to integrate behavior genetic and developmental research approaches to further understand the nature of children's externalizing behavior. First, I review previous research examining genetic and environmental influences on externalizing behavior. Because little research on the genetics of externalizing behavior has used observational methodology, I will present research that has relied on widely used methods: parent-, self-, and teacher-report data, and public record data. This research serves as the foundation upon which observational data can be compared. Next, I present observational data from the Oregon Twin Project (OTP) to describe the genetic and environmental contributions to externalizing behavior during a twin-friend interaction.

GENETICALLY INFORMATIVE RESEARCH ON CHILDREN'S EXTERNALIZING BEHAVIOR

Parent-Report. Behavior genetic researchers generally use parent-report measures to assess externalizing behavior in children. Because parent-report data can be collected via telephone interview, mailed-in questionnaire, or in-person interview, it is the most efficient means of collecting information about children; researchers can collect data from large, geographically diverse samples. Standardized measures, such as the Child Behavior Checklist (CBCL; Achenbach, 1991; Achenbach & Edelbrock, 1983), the Parent Symptom Rating Scale (Conners, 1970), the Behavior Problem Index (Zill, 1985) and the Rutter Antisocial subscale (Rutter, 1967), have shown moderate to substantial genetic variance in aggression when using such methodology (Deater-Deckard, Reiss, Hetherington, & Plomin, 1997; Eley, Lichtenstein, & Stevenson, 1999; Graham & Stevenson, 1985; Hudziak, Rudiger, Neale, Heath & Todd, 2000; Leve et al., 1998; M. O'Connor, Foch, Sherry, & Plomin, 1980; Plomin et al., 1994; Rende et al., 1992; Schmitz et al., 1999; van den Oord, 1996; van den Oord, Boomsma, & Verhulst, 1994; Zahn-Waxler, Schmitz, Fulker, Robinson & Emde, 1996).

However, studies using the same parent-report measure have shown discrepant estimates of genetic and environmental variance (see reviews in Cadoret, Leve, & Devor, 1997; Goldsmith, Gottesman, & Lemery, 1997; Leve et al., 1998). Some parent-report studies find heritability estimates closer to zero (e.g., Schmitz, Cherny, Fulker, & Mrazek, 1994); others show heritability estimates closer to 1 (e.g., Ghodsian-Carpey & Baker, 1987). Potential explanations for such discrepancies include the age of the child, the sample size, the sample type (e.g., at-risk vs. normative or twin vs. adoptive), and the particular subscale or scoring mechanism used. In addition, different forms of externalizing behavior (e.g., oppositional behavior vs. aggression) can show quite different heritability estimates within the same sample (e.g., Eley et al., 1999; Simonoff, Pickles, Meyer, Silberg et al., 1998). Nonetheless, parent-report studies of aggression and antisocial behavior typically show moderate to substantial genetic effects and modest to moderate shared environmental effects.

Self-Report. Similar to parent-report methodology, self-report methodology (with older children and adolescents) provides a cost-effective means of assessing child behavior. In addition, a self-reporting agent *may* have greater insight into their behavior than other reporting agents (e.g., parent-report). Self-report studies of children's externalizing behavior have used standardized measures such as the Delinquency Behavior Inventory (Gibson, 1967) and the Diagnostic Interview Schedule (Robins, Cottler, Bucholz, Compton & Rourke, 1999). Similar to the parent-report findings, most self-report studies of adolescents find that a significant portion of the variance in aggressive and antisocial behav-

ior is attributable to genetic sources, although some studies have also
found significant shared or nonshared variance or both (e.g., Lyons et al.,
1995; Plomin et al., 1994; Rowe, 1983; Rowe, 1986; Rowe, Almeida, &
Jacobson, 1999; Taylor, McGue, Iacono, & Lykken, 2000). As in the parent-report studies, genetic and environmental variance estimates in
self-report studies vary depending on sample variants, leading some to
suggest that there are no consistent patterns of genetic influence on
self-reported aggressiveness (Plomin et al., 1990).

Teacher-Report. Several behavior genetic studies have collected
data from teachers to assess child externalizing behavior. Teachers are
less intimately familiar with the child than parents are, but they have a
broader reference group (i.e., all of the children who have been in their
classroom) by which to compare the focus child. Teachers also observe
children during peer interaction and in a learning context, when the demands on the child differ somewhat from those in the home. Finally, because siblings are typically placed in separate classrooms, each sibling
is often rated by a different teacher, thus reducing the rater bias effects
reported in parent-rating studies (e.g., Hewitt, Silberg, Neale, Eaves &
Erickson, 1992).

Some studies have used measures such as the Teacher Report Form
(Achenbach, 1991) to measure children's externalizing behaviors.
These studies have generally reported modest to moderate genetic effects (Deater-Deckard & Plomin, 1999; Graham & Stevenson, 1985). In
addition, a small adoption study (Jary & Stewart, 1985) measured children's aggressive conduct disorder using five sources (including
teacher-report). Though the effect of teacher-ratings was not isolated,
the study found that, for the adoptive children who were rated as having
aggressive conduct disorder, 30% of their biological mothers and fathers were diagnosed as antisocial; however, none of the adoptive parents of these children were similarly diagnosed, suggesting a genetic
link between child and adult aggressive and antisocial disorders.

Public Records. Public records data, including police arrest records,
Department of Motor Vehicle (DMV) records, and institutional records,
are another means of assessing externalizing behavior in behavior genetic research. These data are particularly useful when studying the biological parent(s) of adopted children, who may not be available for other
types of assessment. Cadoret and colleagues have documented a $G \times E$
interaction for aggression, conduct disorder, and antisocial behavior in
several different samples of adoptees (Cadoret, 1982; Cadoret, Cain, &
Crowe, 1983; Cadoret, Troughton, Bagford, & Woodworth, 1990;
Cadoret, Troughton, & O'Gorman, 1987; Cadoret, Yates, Troughton,

Woodworth, & Stewart, 1995a, 1995b; Jary & Stewart, 1985). The results from these studies suggest that rearing by unstable adoptive parents (as measured by adoptive parent marital discord, disruptive home environment, and/or psychopathology), combined with a genetic risk for antisocial behavior (as measured by birth parent institutional records, police records, or interview measures of antisocial personality disorder and/or substance abuse/dependence), results in multiplicatively higher levels of adoptee aggression and antisocial behavior (as compared to having only one risk factor: an unstable adoptive environment or a genetic risk for antisocial behavior). A Danish adoption study (Cloninger & Gottesman, 1987) provides similar evidence for G × E interactions and genetic main effects using criminality records (Carey, 1992; Cloninger, Sigvardsson, Bohman, & von Knorring, 1982; Gabrielli & Mednick, 1983; Hutchings & Mednick, 1975), as does a Swedish adoption study (Bohman, 1978).

Observation. Because parent-, self- and teacher-report data are susceptible to rater bias in genetic and in nongenetic studies, different sources often show only moderate agreement (e.g., Achenbach, McCanaughy, & Howell, 1987; Bank, Dishion, Skinner, & Patterson, 1990; Simonoff et al., 1995). The correlational differences between monozygotic (MZ) and dizygotic (DZ) twins may in part be a function of biases that individual raters have about the child(ren). For example, Simonoff, Pickles, Meyer et al. (1998) reported striking differences between raters in the relative proportion of genetic and environmental variance in four subtypes of conduct disorder behavior. In light of these rater discrepancies, it is important to examine observational data. Observational data can be less susceptible to rater biases because independent observers code each twin and because the methodology consists of recording the actual behaviors performed rather than asking for an individual's perception of how a child behaves.

A summary of published observational data on the genetics of externalizing behavior is presented in Table 1. In the 10 studies shown, heritability estimates were generally small to modest but varied depending on the sample characteristics and type of observational measure collected. For example, the heritability estimates in the DiLalla and Bishop (1996), the Deater-Deckard (2000), and the Plomin et al. (1981) studies were zero, whereas the Ghodesian-Carpey and Baker (1987) and O'Connor studies (T. O'Connor, Hetherington, Reiss, Plomin, 1995; T. O'Connor, McGuire, Reiss, Hetherington, & Plomin, 1998; T. O'Connor, Neiderhiser, Reiss, Hetherington & Plomin, 1998) demonstrated moderate to substantial genetic effects. The base rate of the be-

TABLE 1. Published Genetically Informative, Observational Data on Externalizing Behavior in Children

Measure	Age	N	h^2	c^2
I. Observational Data				
Global rating of infant negativity with mother. [a]	7-9 months	156	.00	.00
Rate/minute of observed negative action with parents. [b]	2	46	.00	.51
Percentage of observed negative action with parents. [b]	2	46	.14	.36
Percentage of observed commands with parents. [b]	2	46	.14	.61
Global ratings of observed difficult behavior with mother. [c]	3.5-3.75	120	.00	.25
Global ratings of observed child conflict during mother-child-child interaction. [d]	3-11	124	.00	.85
Global ratings of observed child control during mother-child-child interaction. [d]	3-11	124	.00	.29
Frequency of observed negative behavior during unstructured sibling interaction. [d]	3-11	124	.34	.62
Mother-report home observation interval checklist. [e]	4-7	38	.60	.05
Observed frequency of hits to bobo doll. [f]	5-11	87	.00	.42
Observed intensity of hits to bobo doll. [f]	5-11	87	.00	.39
Observed frequency of negative behavior with parent. [g]	6-11	154	.24	.28
Observed frequency of negative behavior controlling for parent behavior. [g]	6-11	154	.30	.14
Global ratings of observed negative behavior with parent. [g]	6-11	154	.29	.27
Global ratings of observed negative behavior controlling for parent behavior. [g]	6-11	154	.58	.00
Global ratings of observed antisocial behavior to mother. [h]	10-18	186	.36	.10
Global ratings of observed antisocial behavior to father. [h]	10-18	186	.31	.17
II. Observational Data Aggregated with Other Measures				
Observation, father-, mother-, and self-report aggregate. [i]	10-18	708	.56	.25
Observation, father-, mother-, and self-report aggregate. [i]	10-18	405	.61	.14

Note. Age is given in years unless otherwise noted; N represents pairs; h^2 = genetic variance; c^2 = shared environmental variance. [a]DiLalla & Bishop (1996); [b]Lytton, Martin, & Eaves (1977); [c]Deater-Deckard (2000); [d]Rende, Slomkowski, Stocker, Fulker, & Plomin (1992); [e]Ghodsian-Carpey & Baker (1987); [f]Plomin, Foch, & Rowe (1981); [g]Leve, Winebarger, Fagot, Reid, & Goldsmith (1998); [h]T. O'Connor, Hetherington, Reiss, & Plomin (1995); [i]T. O'Connor, McGuire, Reiss, Hetherington, & Plomin (1998); [i]T. O'Connor, Neiderhiser, Reiss, Hetherington, & Plomin (1998). For the [b] and [i] studies, h^2 and c^2 (not reported in the original manuscript) were calculated from the MZ and DZ twin correlations. Nonshared environmental variance $(e^2) = 1 - h^2 - c^2$.

havior measured may partially explain the discrepancies; DiLalla and Bishop reported that the baserate for their measure of infant negative behavior was quite low. Age differences in genetic and environmental influences on externalizing behavior may also explain these discrepancies; heritability estimates generally increase as age increases and often do not appear at all until adolescence. Others have hypothesized that genetic effects become stronger as children reach adolescence (e.g., Scarr & McCartney, 1983).

Because each study uses a slightly different observational approach, interpreting differing heritability estimates is difficult–studies may observe the child with the mother, with the co-twin, or by him/herself. In a dyadic interaction, the two participants may influence each other's behavior such that heritability estimates are confounded with participant interaction effects (Leve et al., 1998). Interaction effects could be found if a twin behaved differently (e.g., less hostile) depending on the behaviors of the coparticipant. The studies presented in Table 1 use diverse measurement strategies, including global ratings, rate/minute, percentages or frequencies of behavior, and observation checklists. These methodological differences may contribute to the variability of the genetic and environmental estimates. Overall, however, the genetic variance estimates in these observational studies are considerably lower and the shared environmental estimates considerably higher than in parent-, self-, or teacher-report studies (Emde et al., 1992; Miles & Carey, 1997).

Notably, there are several additional reports from the Nonshared Environment in Adolescent Development project (NEAD; Reiss et al., 2000) that include observational data aggregated with self- and parent-report data (e.g., Reiss et al., 1995). These aggregate studies are noteworthy in their use of a multi-method approach to study genetic and environmental influences on aggressive behavior. The two aggregate studies (reported in the Part II of Table 1) suggest substantial genetic influences and modest to moderate shared and nonshared environmental influences. Additional reports (not presented in Table 1) from aggregated NEAD data have assessed change in antisocial symptoms over time (e.g., Neiderhiser, Reiss, & Hetherington, 1996; T. O'Connor, Neiderhiser et al., 1998) and the covariation between antisocial behavior and parenting practices (e.g., Bussell, Neiderhiser, Pike, Plomin et al., 1999; Neiderhiser et al., 1999; Neiderhiser, Pike, Hetherington, & Reiss, 1998; Pike et al., 1996). An adoption study corroborates the aggregated NEAD results; Jary and Stewart (1985) found genetic effects, environmental effects, and evocative G × E correlations on child antisocial behavior using a composite measure of diagnostic symptoms, and self-, parent, and observer report.

GENETICALLY INFORMATIVE RESEARCH ON CHILDREN'S PEER RELATIONS

Learning about individual differences in the etiology of peer relations is important, as numerous research groups have demonstrated that deviant peer relations are linked to a host of problems for the individual, the family, the school, and the community (e.g., Ary et al., 1999; Dishion et al., 1999; Fergusson, Woodward, & Horwood, 1999; Finnegan, Hodges, & Perry, 1998; Ladd & Burgess, 1999; Loeber, Stouthamer-Loeber, Van Kammen, & Farrington, 1991; Patterson, 1986; Woodward & Fergusson, 1999). Harris (1995) hypothesized peers to be more influential than a child's parents. Additionally, peers may represent an important source of nonshared environmental influences on the child–that is, peer networks and microsocial interactions may differ between members of a twin pair for several reasons. First, because twins are typically placed in separate classrooms from one another, classmates and peer influences are likely to differ. Second, in contrast to their parental relationships, children choose and are chosen by friends. Sibling similarity research suggests that peer relations may differ between siblings in the same family. Daniels and colleagues found that parents and children report only modest sibling similarity in peer friendliness (Daniels, Dunn, Furstenberg, & Plomin, 1985; Daniels & Plomin, 1985).

Using the Sibling Inventory of Differential Experience (SIDE; Daniels & Plomin, 1985), several studies have found evidence for significant genetic influences on self-reported perceptions of peer behavior (Baker & Daniels, 1990; Daniels & Plomin, 1985; Pike, Manke, Reiss & Plomin, 2000). However, the scoring system for the SIDE is problematic for genetic analyses (e.g., Manke, McGuire, Reiss, Hetherington & Plomin, 1995); thus, genetic conclusions drawn from this measure must be interpreted cautiously. Manke et al. used data from the NEAD sample to perform the most thorough investigation of genetic and environmental contributions to children's peer relations to date. The authors collected adolescent self-reports of positive and negative dimensions of interactions with best friends and parent reports of the adolescent's peer group characteristics (college orientation, delinquency, and popularity). The results suggest modest genetic effects for the self-report ratings (ranging from .07 to .31) and moderate to substantial genetic effects for the parent-report ratings (ranging from .49 to .85). Other research investigating the effect that one twin has on the co-twins' externalizing behavior (each twin can be considered the other's peer) suggests that twins can influence each other's externalizing behaviors and that these effects may be stronger for MZs, who share 100% of their familial genes, than for DZs, who share approximately 50% of their familial genes (Carey, 1992; Rowe, 1985; Rowe & Osgood, 1984).

One of the complications in collecting observational measures of peer interaction is that peers may influence one another's behavior. For example, a child who spontaneously yells at their friend may provoke externalizing behavior that may not have transpired otherwise. One approach to resolving this methodological confound is to examine a child's behavior while controlling for the behavior of his or her friend. This study presents twin-friend observation data to demonstrate the effect that a coparticipant can have on interactive behavior.

METHOD

Participants

Participants included twin pairs and a close friend of each twin; all were OTP participants between 1995 and 1996. Twin families had participated in a parent-twin interaction 18 months prior to the twin-friend interaction described here (see Leve et al., 1998). The twins were school-aged, with the median age of 10 years, 2 months (SD = 1 year, 10 months; range = 7-13 years). The sample was comprised of approximately equal numbers of boys and girls. Zygosity was determined with the Zygosity Questionnaire (Goldsmith, 1991), which taps a wide range of developmental and medical history data. Zygosity questionnaires have shown 94% accuracy compared to blood typing (Goldsmith, 1991). The primary caregivers completed the Zygosity Questionnaire, and three independent raters reviewed the questionnaire and photographs of the twins to determine zygosity. In three cases, rater agreement could not be reached; those children were not included in this study. The resulting sample consisted of 76 MZ twin pairs and 72 DZ twin pairs. The DZ twin group included same-sex and opposite-sex DZ twin pairs.

Twin families were identified between 1993 and 1994 via birth announcements, twin organizations, and the public school system in the Willamette Valley of Oregon. Research assistants attempted to contact all eligible twin families. Twin families who were successfully contacted were provided with a description of the study and were invited to participate in the project. Approximately 50% of the families were contacted by phone; the remaining families were contacted by letter. All eligible families who responded to the solicitation were included in the study and were paid for participation.

The twins were primarily European American (91%). Parents ranged from 25 to 54 years of age, with a median 39 years. The mean education level of parents was 15 years. Parent occupational status ranged from 1 to 9 on the Hollingshead occupations code, with a median for fathers equal to 6 (i.e., semiprofessionals and small business owners) and a median for mothers equal to 5 (i.e., clerical and sales workers). Twenty percent of the twins were living in single-parent families. Ethnicity, age of child and parent, parent education, parent occupation, and family structure did not differ by twin zygosity. The demographic background of the twin families was comparable to median levels in the Willamette Valley region from which the sample was drawn.

Procedure

The twins were each asked to select a close friend to participate in the study. The friend was required to be the same age or in the same grade as the twin, to be the same sex as the twin, to be genetically unrelated to the twin, and to be living in a separate household from the twin. Additionally, twins were required to select a different friend than their co-twin had selected. Each twin-friend dyad then participated in two 5-min videotaped discussions at the laboratory. In the first, they were asked to plan a fun activity that they could do together. They were told that it did not have to be anything complicated or expensive, just something that they might enjoy doing and that they thought they could do in the next few weeks. In the second discussion, they were asked to talk about the next school year. They were asked to discuss what it would be like, what they might like about it, what they might not like about it, and what kinds of things they had heard about it from other children.

The order of participation was counterbalanced for the birth order of the twins. Videotapes were coded using the Interpersonal Process Code (IPC; Rusby, Estes, & Dishion, 1991), a microanalytic real-time coding system that simultaneously records three dimensions: *activity, content,* and *affect*. Activity refers to the environment in which the participant is being observed. Content refers to verbal, nonverbal, and physical behavior as it changes through time; there are 24 content codes, which are defined a priori as positive (e.g., *positive interpersonal*), negative (e.g., *physical aggression; coerce*), or neutral (e.g., *neutral command*). Affect refers to the emotional tone accompanying every content code using six ratings: *happy, caring, neutral, distressed, aversive,* and *sad.*

Coder training for the IPC takes approximately 12-16 weeks at 20 hr/week: trainees must achieve 75% event-by-event agreement and a kappa of .65 in two consecutive sessions. Coders then have weekly retraining meetings to maintain reliability and to prevent coder drift. In our study, different research assistants coded each member of any given twin pair. Inter-observer reliability on the IPC was assessed by randomly selecting 15% of the observations to be coded independently by two observers. Inter-rater reliability coefficient alphas were .78 for the content codes (agreement = 90%) and .75 for the affect codes (agreement = 88%).

For each twin and for each friend, we calculated a proportion frequency score for negative behavior: the amount of time that each participant engaged in negative verbal content and/or negative affect during the discussions (on average, once every 30 sec). Thus, negative behavior included talking negatively, performing a negative interpersonal act, being uncooperative, being physically negative, and having an aversive/distressed affect.

Data were entered using a double-entry data file and were analyzed using the augmented DeFries-Fulker (DF) regression approach for behavioral genetic model fitting (Cyphers, Phillips, Fulker, & Mrazek, 1990; DeFries & Fulker, 1985); this allows for simultaneous testing and estimation of genetic and shared environmental influences. See Petrill (this volume) for a more complete description of this methodological approach.

RESULTS

Descriptive Information

The *M* and *SD* information is presented in Table 2. To preclude potential differences owing to the age or sex of the twins and their friends (especially because the sample included same-sex and opposite-sex DZ twins), Age, Sex, and Age × Sex were regressed out of the remaining analyses.

The correlation between twin negative behavior and friend negative behavior was .63 (MZ = .75; DZ = .50). The raw intraclass correlations and the correlations adjusted for Age, Sex, and Age × Sex are presented in Table 3, Part I; the correlations controlling for the coparticipants behavior (described later in Results) are presented in Table 3, Part II. In most cases, Age, Sex, and Age × Sex adjustments made little difference on the relative magnitude of the correlations.

As is shown in Part I of the table, there are some differences between the MZ and DZ correlations. The MZ-DZ correlations for twin negative behavior suggest moderate genetic effects, and the correlations for friend negative behavior suggest modest environmental effects. The low intra-twin and intra-friend correlations also suggest substantial

TABLE 2. Means and Standard Deviations for the Twin-Friend Negative Behavior Scores

	MZ		DZ	
Measures	M	SD	M	SD
Twin Negative Behavior	0.52	0.56	0.53	0.58
Friend Negative Behavior	0.54	0.61	0.52	0.59

Note. There were no significant mean-level or variance differences by zygosity.

TABLE 3. Raw and Adjusted Intraclass Correlations as Measures of Twin Similarity

	MZ		DZ	
Measures	Raw r (SE)[b]	Adjusted r[a] (SE)[a]	Raw r (SE)[b]	Adjusted r[a] (SE)[b]
I. Twin-Friend Negative Behavior Scores				
Twin Negative Behavior	.25 (.11)	.23 (.11)	−.05 (.12)	−.03 (.12)
Friend Negative Behavior	.14 (.11)	.12 (.12)	.17 (.12)	.12 (.12)
II. Negative Behavior Scores with Coparticipants Behavior Partialed Out				
Twin Negative Behavior	−.03 (.12)	.03 (.12)	−.10 (.12)	−.10 (.12)
Friend Negative Behavior	−.16 (.11)	−.15 (.11)	.07 (.12)	.02 (.12)

[a] Corrected for Age, Sex, and Age × Sex interaction differences
[b] Corrected for the appropriate degrees of freedom

nonshared environmental effects. By subtracting the MZ correlation from 1 (Falconer, 1989) and using the Age, Sex, and Age × Sex adjusted correlations, the genetic estimates for the twin negative and friend negative behavior are .23 and .00, respectively; nonshared environment estimates are .77 and .88, respectively.

Model Fitting

Next, we conducted a series of DF regression models; the best-fitting models are presented in Table 4 (Part II contains the regression results with the co-participant's behavior partialed out). These models were developed following the DF regression procedures outlined in Eley (this volume) and Petrill (this volume).

As is shown in Part I of Table 4, dominant genetic effects were significant for twin negative behavior, and neither genetic nor shared environmental effects were significant for friend negative behavior. Nonshared environmental variance appears to contribute to twin behavior and to friend behavior, as shown by the interpolated E parameter. Measurement error also plays a role.

It is possible that some of the variation in each individual's behavior is the direct result of the coparticipant's behavior. As presented in Leve et al. (1998), coparticipants can directly affect each other's behavior.

TABLE 4. DeFries-Fulker Regression Results for Twin-Friend Negative Behavior Scores

Measure	Model	A	D	C	E
I. Twin-Friend Negative Behavior					
Twin Negative Behavior	DE	-	.15* (.11)	-	.85
Friend Negative Behavior	CE	-	-	.12	.88
II. Negative Behavior with Coparticipants Behavior Partialed Out					
Twin Negative Behavior	E	-	-	-	1.00
Friend Negative Behavior	E	-	-	-	1.00

Note. Numbers in parentheses are the standard errors, corrected for the appropriate degrees of freedom. A = additive genetic variance; D = dominant (nonadditive) genetic variance; C = shared environmental variance; E = nonshared environmental variance.
*$p < .05$.

We hypothesized that our sample could have produced such variation. For example, a particularly aggressive friend may elicit externalizing behaviors by the twin. To better understand the extent to which gene-environment processes can be modified by interactions with others, we conducted a second series of regression models that was free from the effects of the coparticipant's behavior. Although such analyses present statistically altered data, they increase our ability to identify specific sources of environmental variance. Part II of Table 3 provides the raw and adjusted correlations for the interaction variables with the coparticipant's behavior partialed out. Compared to Part 1 of Table 3, these correlations were reduced to minor fluctuations around zero. The remaining variability in the twin and friend negative behavior ($SD = .44$ as compared to the original $SD = .53$) suggested that the reduction of the correlations did not result from a reduction in the negative behavior variability. Rather, they imply that there was no correspondence between what co-twins did (in separate interactions), when controlling for the friends' behavior.

The best-fitting DF regression models with the coparticipant's behavior partialed out are presented in Part II of Table 4. These models analyzed the residual score after the coparticipant's behavior had been controlled for; for example, friend behavior was partialed out of the twin's behavior when predicting to twin negative behavior. Both of the residualized models suggested a nonshared environment model; the A, C, and D parameters were all negative, suggesting negligible effects of genes and shared environment. The results from the nonresidualized models (see Part I) also suggested strong nonshared environmental effects.

DISCUSSION

In contrast to the moderate twin similarity observed during a parent-child interaction task (reported in Leve et al., 1998), twin (and friend) similarity during the twin-friend discussion was modest to nonexistent. Compared to the DZ correlations, the MZ twin correlations were larger and the MZ friend correlations were similar in magnitude. The DF regressions models (without controlling for the coparticipant's behavior) suggested that dominant genetic factors contributed to twin negative behavior and that shared environmental factors contributed to friend negative behavior. The interpolated, nonshared environmental estimates were substantial for both measures. Although there are no genetically informative observational studies of externalizing peer behavior to compare these results to, a questionnaire study of twins' peer behavior also suggested a primary role for nonshared environmental influences on peer behavior when self-report methods were used (Manke et al., 1995). However, the same study found a primary role for genetic influences on peer behavior when parent-report methods were used.

One explanation for the dominant genetic effects contributing to twin negative behavior during the friend interaction involves a greater similarity of MZ twins' friends than DZ twins' friends. Our examination of the friend-friend correlations (.12 for MZ and for DZ) suggested otherwise, thus supporting a fundamental assumption in the twin design, the equal environment assumption (EEA). The EEA states that environmental influences should not differ across sibling pairs as a function of genetic relatedness.

However, the MZ twin-friend correlation was higher than the DZ twin-friend correlation (.75 vs. .50), illustrating that the MZ twins resembled their friends more than the DZ twins resembled their friends. Having grown up with a genetically identical sibling, MZ twins may be more accustomed to being intimate with someone who is similar to them. Further research is needed to investigate this question and its long-term implications. It would be particularly useful to test whether this correlational pattern exists for twins who are reared apart.

With coparticipants' behavior partialed out of the twin's/friend's behavior, the DF regression models suggested only nonshared environmental effects. Although several models with differing variance components were tested, the genetic and shared environmental influences were near zero, as was expected given that the MZ and DZ correlations became negative or close to zero when controlling for the coparticipant's behavior. This suggests that genetic effects on behavior can be elicited when interacting with a friend. However, with the friend's negative behavior partialed out, nonshared sources influence twins' behavior, as was expected given the methodological design of this study; each twin participated with a different friend than his or her co-twin. As was hypothesized, this study found that friends represent a source of nonshared environmental variation on twins' externalizing behavior. The question remains as to whether the twin-friend interaction behavior is typical of the twin's behavior in other contexts.

Why Do Observational Studies Obtain Different Estimates than Self- or Parent-Report Studies?

In general, observational data produce lower intratwin pair correlations than do self- or parent-report data (e.g., Reiss et al., 1995). We found similar trends in this study. The decreased twin-pair correlations had an effect on the basic formulas for computing genetic and environmental contributions to behavior. Because it is computed as how MZ twins differ from one another ($1 - r_{mz}$), nonshared environment is likely to appear greater in observational studies (vs. self- or parent-report studies) simply because the intratwin pair correlations are smaller.

Additionally, compared to independent observers, parents may be more likely to report on their child's behavior without reference to what the norms are for this age of child; parents may only compare and contrast each twin to his/her co-twin or with other siblings in the family. This type of rater bias has the potential to skew estimates of genetic and environmental variance. For example, if the aggression levels of a pair of MZ twins are contrasted against one another, this correlation and the ensuing genetic estimate may be artificially low, whereas the nonshared environmental variance may be artificially high (e.g., Eaves & Carbonneau, 1998; Simonoff, Pickles, Hervas, Silberg et al., 1998; Simonoff, Pickles, Meyer et al., 1998). Conversely, if a parent exaggerates or overestimates the similarity of aggression in a pair of MZ twins (perhaps believing that MZ twins are supposed to be similar to one another), the resulting correlation and genetic estimate may be artificially high. Using the CBCL, some researchers have found rater bias effects (e.g., Graham & Stevenson, 1985; Hewitt et al., 1992; Simonoff et al., 1995), while others have not (Hudziak et al., 2000).

Furthermore, the specific observational contexts examined in this study may have produced the greater nonshared environmental effects and the lower genetic effects compared to questionnaire studies. Within this sample, the genetic and environmental components of externalizing behavior were influenced by the observational setting. Although the parent-child interactions reported in Leve et al. (1998) and the twin-friend interactions reported here both used observationally coded measures of twin negative behavior, the estimates for the two observational contexts appear quite different. Shared environmental and genetic variance was larger in the parent-child interaction context than in the twin-friend interaction context, whereas nonshared environment was larger in the twin-friend context (although significance testing between the studies was not conducted). This suggests that the etiology of externalizing behavior may differ depending on the setting, the type, and the form of the externalizing behavior. Different intervention techniques may thus be more effective depending on the targeted setting.

Limitations and Future Directions

One limitation of this research is that our methodology focused on genetic and environmental effects in isolation from one another; increasing evidence suggests that genes and the environment do not operate independently (e.g., Bergeman, Plomin, Pedersen, & McClearn, 1991; Neiderhiser et al., 1999; Pike et al., 1996). Ge et al. (1996) found that adopted children at genetic risk for antisocial behavior or substance abuse/dependency elicited more negative parenting from their mothers and fathers than adoptive children not at genetic risk. Similarly, T. O'Connor, Deater-Deckard, Fulker, Rutter and Plomin (1998) found that adopted children at genetic risk for antisocial behavior were more likely (than adopted children not at genetic risk) to elicit negative parenting. The Ge and the O'Connor findings each describe an evocative G × E correlation in which environmental and genetic effects are not independent. Future research should further examine how genetic and environmental effects separately contribute to behavior and how they mutually influence externalizing outcomes.

As a note of caution, heritability estimates from twin designs are often higher than those from adoption designs. In a study of 88 adoptive 7- to 12-year-old children, T. O'Connor, Deater-Deckard et al. (1998) were unable to find a significant genetic influence on children's externalizing behavior on the Externalizing factor, the Aggressiveness scale, or the Delinquent Behavior scale of the CBCL. This stands in sharp contrast to parent-report studies of the CBCL in twin samples. Observational research on the etiology of aggressive and antisocial behavior in adoptive families is needed to extend the research findings presented here and to test whether they hold in adoptive samples. Additionally, there is evidence to suggest that genetic contributions to externalizing behavior may differ between boys and girls (e.g., Eley et al., 1999; Rodgers, Rowe, & Buster, 1999; Rowe, 1985), although other studies report no sex differences (e.g., Schmitz et al., 1994). Unfortunately, the OTP sample was too small to permit analyses of sex differences in the etiology of externalizing behavior. Further research examining sex differences in observational studies of externalizing behavior is needed to extend this research.

Lastly, although our sample is comparable in size to other developmental observation studies and many behavior genetic studies (e.g., Deater-Deckard, 2000; Deater-Deckard & Plomin, 1999; Fagot, 1997; Ghodsian-Carpey & Baker, 1987; Shaw, Keenen, & Vondra, 1994), it is smaller than questionnaire and interview twin studies from population-based research (e.g., Kendler, Neale, Kessler, Heath & Eaves, 1992; Silberg, Rutter, Meyer, & Maes, 1996). It is logistically difficult and prohibitively expensive to conduct a large-scale observational study comparable in size to the population-based twin studies. With this smaller sample, however, we have been able to measure parent-child (Leve et al., 1998) and twin-friend interactive behavior, thereby addressing several unresolved questions concerning the role of assessment methodology in the estimation of genetic and environmental influences on externalizing behavior. Future observational research is needed to build upon the data presented here and to entertain more elaborate hypotheses about the role of method variation in genetic studies.

NOTE

1. Throughout this manuscript, we use the term "externalizing behavior" as a general term to include the class of behaviors such as aggression, antisocial behavior, hostility, disobedience, fighting, stealing, acting out, and lying.

REFERENCES

Achenbach, T. M. (1991). *Manual for the Child Behavior Checklist/4-18 and 1991 Profile*. Burlington: University of Vermont, Department of Psychiatry.

Achenbach, T. M., & Edelbrock, C. (1983). *Manual for the Child Behavior Checklist and Revised Child Behavior Profile*. Burlington: University of Vermont, Department of Psychiatry.

Achenbach, T. M., McConaughy, S. H., & Howell, C. T. (1987). Child/adolescent behavioral and emotional problems: Implications of cross-informant correlations for situational specificity. *Psychological Bulletin, 101*, 213-232.

Ary, D. V., Duncan, T. E., Biglan, A., Metzler, C. W., Noell, J. W., & Smolkowski, K. (1999). Development of adolescent problem behavior. *Journal of Abnormal Child Psychology, 27*(2), 141-150.

Baker, L. A., & Daniels, D. (1990). Nonshared environmental influences and personality differences in adult twins. *Journal of Personality and Social Psychology, 58*(1), 103-110.

Bank, L., Dishion, T., Skinner, M., & Patterson, G. R. (1990). Method variance in structural equation modeling: Living with "Glop." In G. R. Patterson (Ed.), *Depression and aggression in family interaction*. Hillsdale: Lawrence Erlbaum.

Bergeman, C. S., Plomin, R., Pedersen, N. L., & McClearn, G. E. (1991). Genetic mediation of the relationship between social support and psychological well-being. *Psychology and Aging, 6*(4), 640-646.

Bohman, M. (1978). Some genetic aspects of alcoholism and criminality. *Archives of General Psychiatry, 35,* 269-276.

Brook, J. S., Whiteman, M., Finch, S. J., & Cohen, P. (1996). Young adult drug use and delinquency: Childhood antecedents and adolescent mediators. *Journal of the American Academy of Child and Adolescent Psychiatry, 35*(12), 1584-1592.

Bussell, D. A., Neiderhiser, J. M., Pike, A., Plomin, R., Simmens, S., Howe, G. W. et al. (1999). Adolescents' relationships to siblings and mothers: A multivariate genetic analysis. *Developmental Psychology, 35*(5), 1248-1259.

Cadoret, R. J. (1982). Genotype-environment interaction in antisocial behaviour. *Psychological Medicine, 12,* 235-239.

Cadoret, R. J., Cain, C. A., & Crowe, R. R. (1983). Evidence for gene-environment interaction in the development of adolescent antisocial behavior. *Behavior Genetics, 13*(3), 301-310.

Cadoret, R. J., Leve, L. D., & Devor, E. (1997). Genetics of aggressive and violent behavior. *Psychiatric Clinics of North America, 20*(2), 301-322.

Cadoret, R. J., Troughton, E., Bagford, J., & Woodworth, G. (1990). Genetic and environmental factors in adoptee antisocial personality. *European Archives of Psychiatry and Neurological Sciences, 239,* 231-240.

Cadoret, R. J., Troughton, E., & O'Gorman, T. W. (1987). Genetic and environmental factors in alcohol abuse and antisocial personality. *Journal of Studies on Alcohol, 48*(1), 1-8.

Cadoret, R. J., Yates, W. R., Troughton, E., Woodworth, G., & Stewart, M. A. (1995a). Genetic-environmental interaction in the genesis of aggressivity and conduct disorders. *Archives of General Psychiatry, 52,* 916-924.

Cadoret, R. J., Yates, W. R., Troughton, E., Wooworth, G., & Stewart, M. A. (1995b). Adoption study demonstrating two genetic pathways to drug abuse. *Archives of General Psychiatry, 52,* 42-52.

Carey, G. (1992). Twin imitation for antisocial behavior: Implications for genetic and family environment research. *Journal of Abnormal Psychology, 101*(1), 18-25.

Cloninger, C. R., & Gottesman, I. I. (1987). Genetic and environmental factors in antisocial behavior disorders. In S. A. Mednick, T. E. Moffitt, & S. A. Stack (Eds.), *The causes of crime: New biological approaches* (pp. 92-109). New York: Cambridge University Press.

Cloninger, C. R., Sigvardsson, S., Bohman, M., & von Knorring, A.L. (1982). Predisposition to petty criminality in Swedish adoptees. *Archives of General Psychiatry, 39*(2), 1242-1247.

Conners, C. K. (1970). Symptom patterns in hyperkinetic, neurotic, and normal children. *Child Development, 41,* 667-682.

Cyphers, L. A., Phillips, K., Fulker, D. W., & Mrazek, D. A. (1990). Twin temperament during the transition from infancy to early childhood. *Journal of the American Academy of Child and Adolescent Psychiatry, 29*, 393-397.

Daniels, D., Dunn, J., Furstenberg, F. F., Jr., & Plomin, R. (1985). Environmental differences within the family and adjustment differences within pairs of adolescent siblings. *Child Development, 56*, 764-774.

Daniels, D., & Plomin, R. (1985). Differential experience of siblings in the same family. *Developmental Psychology, 21*(5), 747-760.

Deater-Deckard, K. (2000). Parenting and child behavioral adjustment in early childhood: A quantitative genetic approach to studying family processes. *Child Development, 71*(2), 468-484.

Deater-Deckard, K., Dodge, K. A., Bates, J. E., & Pettit, G. S. (1998). Multiple risk factors in the development of externalizing behavior problems: Group and individual differences. *Development and Psychopathology, 10*(3), 469-493.

Deater-Deckard, K., & Plomin, R. (1999). An adoption study of the etiology of teacher and parent reports of externalizing behavior problems in middle childhood. *Child Development, 70*(1), 144-154.

Deater-Deckard, K., Reiss, D., Hetherington, E. M., & Plomin, R. (1997). Dimensions and disorders of adolescent adjustment: A quantitative genetic analysis of unselected samples and selected extremes. *Journal of Child Psychology and Psychiatry and Allied Disciplines, 38*(5), 515-525.

DeFries, J. C., & Fulker, D. W. (1985). Multiple regression analysis of twin data. *Behavior Genetics, 15*(5), 467-473.

DiLalla, L. F., & Bishop, E. G. (1996). Differential maternal treatment of infant twins: Effects on infant behaviors. *Behavior Genetics, 26*(6), 535-542.

Dishion, T. J., McCord, J., & Poulin, F. (1999). When interventions harm: Peer groups and problem behavior. *American Psychologist, 54*(9), 755-764.

Eaves, L. J., & Carbonneau, R. (1998). Recovering components of variance from differential ratings of behavior and environment in pairs of relatives. *Developmental Psychology, 34*(1), 125-129.

Eley, T. C., Lichtenstein, P., & Stevenson, J. (1999). Sex differences in the etiology of aggressive and nonaggressive antisocial behavior: Results from two twin studies. *Child Development, 70*(1), 155-168.

Emde, R. N., Plomin, R., Robinson, J., Corley, R., DeFries, J., Fulker, D. W., Reznick, J. S., Campos, J., Kagan, J., & Zahn-Waxler, C. (1992). Temperament, emotion, and cognition at fourteen months: The MacArthur Longitudinal Twin Study. *Child Development, 63*, 1437-1455.

Fagot, B. I. (1997). Attachment, parenting, and peer interaction of toddler children. *Developmental Psychology, 33*, 489-499.

Fagot, B. I., & Leve, L. D. (1998). Teacher ratings of externalizing behavior at school entry for boys and girls: Similar early predictors and different correlates. *Journal of Child Psychology and Psychiatry and Allied Disciplines, 39*(4), 555-566.

Falconer, D. S. (1989). *Introduction to quantitative genetics* (3rd ed.). New York: Longman.

Fergusson, D. M., Horwood, L. J., & Lynskey, M. (1994). The childhoods of multiple problem adolescents: A 15-year longitudinal study. *Journal of Child Psychology and Psychiatry*, *35*(6), 1123-1140.

Fergusson, D. M., Woodward, L. J., & Horwood, L. J. (1999). Childhood peer relationship problems and young people's involvement with deviant peers in adolescence. *Journal of Abnormal Child Psychology*, *27*(5), 357-370.

Finnegan, R. A., Hodges, E. V. E., & Perry, D. G. (1998). Victimization by peers: Associations with children's reports of mother-child interaction. *Journal of Personality and Social Psychology*, *75*(4), 1076-1086.

Gabrielli, W. F., Jr., & Mednick, S. A. (1983). Genetic correlates of criminal behavior. *American Behavioral Scientist*, *27*(1), 59-74.

Ge, X., Conger, R. D., Cadoret, R. J., Neiderhiser, J. M., Yates, W., Troughton, E. et al. (1996). The developmental interface between nature and nurture: A mutual influence model of child antisocial behavior and parent behaviors. *Developmental Psychology*, *32*, 574-589.

Ghodsian-Carpey, J., & Baker, L. A. (1987). Genetic and environmental influences on aggression in 4- to 7-year-old twins. *Aggressive Behavior*, *13*, 173-186.

Gibson, H. B. (1967). Self-reported delinquency among schoolboys, and their attitudes to the police. *British Journal of Social and Clinical Psychology*, *6*, 168-173.

Goldsmith, H. H. (1991). A zygosity questionnaire for young twins: A research note. *Behavior Genetics*, *214*, 257-269.

Goldsmith, H. H., Gottesman, I. I., & Lemery, K. S. (1997). Epigenetic approaches to developmental psychopathology. *Development and Psychopathology*, *94*, 365-387.

Graham, P., & Stevenson, J. (1985). A twin study of genetic influences on behavioral deviance. *American Academy of Child Psychiatry*, *24*(1), 33-41.

Harris, J. A. (1995). Where is the child's environment? A group socialization theory of development. *Psychological Review*, *102*, 458-489.

Hewitt, J. K., Silberg, J. L., Neale, M. C., Eaves, L. J., & Erickson, M. (1992). The analysis of parental ratings of children's behavior using LISREL. *Behavior Genetics*, *22*(3), 293-317.

Hudziak, J. J., Rudiger, L. P., Neale, M. C., Heath, A. C., & Todd, R. D. (2000). A twin study of inattentive, aggressive, and anxious/depressed behaviors. *Journal of the American Academy of Child and Adolescent Psychiatry*, *39*(4), 469-476.

Hutchings, B., & Mednick, S. A. (1975). Registered criminality in the adoptive and biological parents of registered male criminal adoptees. In R. R. Fieve, D. Rosenthal, & H. Brill (Eds.), *Genetic research in psychiatry* (pp. 105-116). Baltimore: Johns Hopkins University Press.

Jary, M. L., & Stewart, M. A. (1985). Psychiatric disorder in the parents of adopted children with aggressive conduct disorder. *Neuropsychobiology*, *13*, 7-11.

Keenan, K., & Shaw, D. S. (1995). The development of coercive family processes: The interaction between aversive toddler behavior and parenting factors. In J. McCord (Ed.), *Coercion and punishment in long-term perspectives* (pp. 165-180). Cambridge, UK: Cambridge University Press.

Kendler, K. S., Neale, M. C., Kessler, R. C., Heath, A. C., & Eaves, L. J. (1992). A population based twin study of major depression in women: The impact of varying definitions of illness. *Archives of General Psychiatry*, *49*, 257-266.

Kim, J. E., Hetherington, E. M., & Reiss, D. (1999). Associations among family relationships, antisocial peers, and adolescents' externalizing behaviors: Gender and family type differences. *Child Development, 70*(5), 1209-1230.

Ladd, G. W., & Burgess, K. B. (1999). Charting the relationship trajectories of aggressive, withdrawn, and aggressive/withdrawn children during early grade school. *Child Development, 70*(4), 910-929.

Leve, L. D., Winebarger, A. A., Fagot, B. I., Reid, J. B., & Goldsmith, H. H. (1998). Environmental and genetic variance in children's observed and reported maladaptive behavior. *Child Development, 69*(5), 1286-1298.

Loeber, R., & Dishion, T. (1983). Early predictors of male delinquency: A review. *Psychological Bulletin, 94*(1), 68-99.

Loeber, R., Stouthamer-Loeber, M., Van Kammen, W., & Farrington, D. P. (1991). Initiation, escalation and desistance in juvenile offending and their correlates. *The Journal of Criminal Law and Criminology, 82*(1), 36-82.

Lyons, M. J., True, W. R., Eisen, S. A., Goldberg, J., Meyer, J. M., Faraone, S. V. et al. (1995). Differential heritability of adult and juvenile antisocial traits. *Archives of General Psychiatry, 52*, 906-915.

Lytton, H., Martin, G., & Eaves, L. (1977). Environmental and genetical causes of variation in ethological aspects of behavior in two-year-old boys. *Social Biology, 24*, 200-210.

Maccoby, E. E. (2000). Parenting and its effects on children: On reading and misreading behavior genetics. *Annual Review of Psychology, 51*, 1-27.

Manke, B., McGuire, S., Reiss, D., Hetherington, E. M., & Plomin, R. (1995). Genetic contributions to adolescents' extrafamilial social interactions: Teachers, best friends, and peers. *Social Development, 4*(3), 238-256.

McMahon, R. J. (1994). Diagnosis, assessment, and treatment of externalizing problems in children: The role of longitudinal data. *Journal of Consulting and Clinical Psychology, 62*(5), 901-917.

Miles, D. R., & Carey, G. (1997). Genetic and environmental architecture of human aggression. *Journal of Personality and Social Psychology, 7*(1), 207-217.

Neale, M. C., & Cardon, L. R. (1992). *Methodology for genetic studies of twins and families.* Dordrecht, The Netherlands: Kluwer Academic Publishers.

Neiderhiser, J. M., Pike, A., Hetherington, E. M., & Reiss, D. (1998). Adolescent perceptions as mediators of parenting: Genetic and environmental contributions. *Developmental Psychology, 34*(6), 1459-1469.

Neiderhiser, J. M., Reiss, D., & Hetherington, M. E. (1996). Genetically informative designs for distinguishing developmental pathways during adolescence: Responsible and antisocial behavior. *Development and Psychopathology, 8*, 779-791.

Neiderhiser, J. M., Reiss, D., Hetherington, E. M., & Plomin, R. (1999). Relationships between parenting and adolescent adjustment over time: Genetic and environmental contributions. *Developmental Psychology, 35*(3), 680-692.

O'Connor, M., Foch, T., Sherry, T., & Plomin, R. (1980). A twin study of specific behavioral problems of socialization as viewed by parents. *Journal of Abnormal Child Psychology, 8*(2), 189-199.

O'Connor, T. G., Deater-Deckard, K., Fulker, D., Rutter, M., & Plomin, R. (1998). Genotype-environment correlations in late childhood and early adolescence: Antisocial behavioral problems and coercive parenting. *Developmental Psychology, 34*(5), 970-981.

O'Connor, T. G., Hetherington, E. M., Reiss, D., & Plomin, R. (1995). A twin-sibling study of observed parent-adolescent interactions. *Child Development, 66,* 812-829.

O'Connor, T. G., McGuire, S., Reiss, D., Hetherington, E. M., & Plomin, R. (1998). Co-occurrence of depressive symptoms and antisocial behavior in adolescence: A common genetic liability. *Journal of Abnormal Psychology, 107*(1), 27-37.

O'Connor, T. G., Neiderhiser, J. M., Reiss, D., Hetherington, E. M., & Plomin, R. (1998). Genetic contributions to continuity, change, and co-occurrence of antisocial and depressive symptoms in adolescence. *Journal of Child Psychology and Psychiatry and Allied Disciplines, 39*(3), 323-336.

Patterson, G. R. (1976). The aggressive child: Victim and architect of a coercive system. In E. J. Marsh, L. A. Hamerlynck, & L. Handy (Eds.), *Behavior modification and families* (pp. 267-316). New York: Brunner/Mazel.

Patterson, G. R. (1982). *Coercive family processes.* Eugene, OR: Castalia.

Patterson, G. R. (1986). Performance models for antisocial boys. *American Psychologist, 41*(4), 432-444.

Patterson, G. R., Capaldi, D., & Bank, L. (1991). An early starter model for predicting delinquency. In D. J. Pepler & K. H. Rubin (Eds.), *The development and treatment of childhood aggression* (pp. 139-168). Hillsdale: Lawrence Erlbaum.

Patterson, G. R., Reid, J. B., & Dishion, T. J. (1992). *Antisocial boys.* Eugene, OR: Castalia.

Pike, A., Manke, B., Reiss, D., & Plomin, R. (2000). A genetic analysis of differential experiences of adolescent siblings across three years. *Social Development, 9*(1), 96-114.

Pike, A. McGuire, S., Hetherington, E. M., Reiss, D., & Plomin, R. (1996). Family environment and adolescent depressive symptoms and antisocial behavior: A multivariate genetic analysis. *Developmental Psychology, 32*(4), 590-603.

Plomin, R., Foch, T. T., & Rowe, D. C. (1981). Bobo clown aggression in childhood: Environment not genes. *Journal of Research in Personality, 15,* 331-342.

Plomin, R., Nitz, K., & Rowe, D. C. (1990). Behavioral genetics and aggressive behavior in childhood. In M. Lewis & S. M. Miller (Eds.), *Handbook of developmental psychopathology* (pp. 119-133). New York: Plenum Press.

Plomin, R., Reiss, D., Hetherington, E. M., & Howe, G. W. (1994). Nature and nurture: Genetic contributions to measures of the family environment. *Developmental Psychology, 30*(1), 32-43.

Reiss, D., Hetherington, E. M., Plomin, R., Howe, G. W., Simmens, S. J., Henderson, S. H. et al. (1995). Genetic questions for environmental studies: Differential parenting and psychopathology in adolescence. *Archives of General Psychiatry, 52,* 925-936.

Reiss, D., Neiderhiser, J. M., Hetherington, E. M., & Plomin, R. (2000). *The relationship code: Deciphering genetic and social influences on adolescent development.* Cambridge, MA: Harvard University Press.

Rende, R. D., Slomkowski, C. L., Stocker, C., Fulker, D. W., & Plomin, R. (1992). Genetic and environmental influences on maternal and sibling interaction in middle childhood: A sibling adoption study. *Developmental Psychology, 28*(3), 484-490.

Robins, L., Cottler, L, Bucholz, K., Compton, W., & Rourke, K. M. (1999). *Diagnostic Interview Schedule for DSM IV (DIS-IV)*. Available from the authors.

Rodgers, J. L., Rowe, D. C., & Buster, M. (1999). Nature, nurture and first sexual intercourse in the USA: Fitting behavioural genetic models to NLSY kinship data. *Journal of Biosocial Science, 31*(1), 29-41.

Rowe, D. C. (1983). Biometrical genetic models of self-reported delinquent behavior: A twin study. *Behavior Genetics, 13*(5), 473-489.

Rowe, D. C. (1985). Sibling interaction and self-reported delinquent behavior: A study of 265 twin pairs. *Criminology, 23*(2), 223-239.

Rowe, D. C. (1986). Genetic and environmental components of antisocial behavior: A study of 265 twin pairs. *Criminology, 24*(3), 513-532.

Rowe, D. C., Almeida, D. M., & Jacobson, K. C. (1999). School context and genetic influences on aggression in adolescence. *Psychological Science, 10*(3), 277-280.

Rowe, D. C., & Osgood, D. W. (1984). Heredity and sociological theories of delinquency: A reconsideration. *American Sociological Review, 49*(August), 526-540.

Rusby, J. C., Estes, A., & Dishion, T. (1991). *The Interpersonal Process Code (IPC)*. (Available from Oregon Social Learning Center, 160 East 4th Avenue, Eugene, OR 97401-2426).

Rutter, M. (1967). A children's behavior questionnaire for completion by teachers: Preliminary findings. *Journal of Child Psychology and Psychiatry and Allied Disciplines, 22*, 323-356.

Rutter, M. L. (1997). Nature–nurture integration. *American Psychologist, 52*(4), 390-398.

Scarr, S., & McCartney, K. (1983). How people make their own environments: A theory of genotype -> environment effects. *Child Development, 54*, 424-435.

Schmitz, S., Cherny, S. S., Fulker, D. W., & Mrazek, D. A. (1994). Genetic and environmental influences on early childhood behavior. *Behavior Genetics, 24*(1), 25-34.

Schmitz, S., Fulker, D. W., Plomin, R., Zahn-Waxler, C., Emde, R. N., & DeFries, J. C. (1999). Temperament and problem behaviour during early childhood. *International Journal of Behavioral Development, 23*(2), 333-355.

Shaw, D. S., Keenan, K., & Vondra, J. I. (1994). Developmental precursors of externalizing behavior: Ages 1 to 3. *Developmental Psychology, 30*, 355-364.

Silberg, J. L., Rutter, M., Meyer, J., & Maes, H. (1996). Genetic and environmental influences on covariation between hyperactivity and conduct disturbance in juvenile twins. *Journal of Child Psychology and Psychiatry and Allied Disciplines, 37*, 803-816.

Simonoff, E., Pickles, A., Hewitt, J., Silberg, J., Rutter, M., Loeber, R. et al. (1995). Multiple raters of disruptive child behavior: Using a genetic strategy to examine shared views and bias. *Behavior Genetics, 25*(4), 311-326.

Simonoff, E., Pickles, A., Hervas, A., Silberg, J. L., Rutter, M., & Eaves, L. (1998). Genetic influences on childhood hyperactivity: Contrast effects imply parental rating bias, not sibling interaction. *Psychological Medicine, 28*(4) 825-837.

Simonoff, E., Pickles, A., Meyer, J., Silberg, J., & Maes, H. (1998). Genetic and environmental influences on subtypes of conduct disorder behavior in boys. *Journal of Abnormal Child Psychology, 26*(6), 495-509.

Stoolmiller, M., Patterson, G. R., & Snyder, J. (1997). Parental discipline and child antisocial behavior: A contingency-based theory and some methodological refinements. *Psychological Inquiry, 8*(3), 223-229.

Taylor, J., McGue, M., Iacono, W. G., & Lykken, D. T. (2000). A behavioral genetic analysis of the relationship between the socialization scale and self-reported delinquency. *Journal of Personality, 68*(1), 29-50.

van den Oord, E. J. (1996). Problem behaviours: Genetic & environmental influences. *Journal of Abnormal Psychology, 105*(3), 349-357.

van den Oord, E. J., Boomsma, D. I., & Verhulst, F. C. (1994). A study of problem behaviors in 10- to 15-year-old biologically related and unrelated international adoptees. *Behavior Genetics, 24*(3), 193-205.

Vitaro, F., Brendgen, M., Pagani, L., Tremblay, R. E., & McDuff, P. (1999). Disruptive behavior, peer association, and conduct disorder: Testing the developmental links through early intervention. *Development and Psychopathology, 11,* 287-304.

Woodward, L. J., & Fergusson, D. M. (1999). Childhood peer relationship problems and psychosocial adjustment in late adolescence. *Journal of Abnormal Child Psychology, 27*(1), 87-104.

Zahn-Waxler, C., Schmitz, S., Fulker, D., Robinson, J., & Emde, R. (1996). Behavior problems in 5-year-old monozygotic and dizygotic twins: Genetic and environmental influences, patterns of regulation, and internalization of control. *Development and Psychopathology, 8,* 103-122.

Zill, N. (1985). *Behaviour problems scale developed for the 1981 Child Health supplement to the National Health Interview Survey.* Washington, DC: Child Trends.

Longitudinal Connections Between Parenting and Peer Relationships in Adoptive and Biological Families

Thomas G. O'Connor
Jennifer M. Jenkins
John Hewitt
John C. DeFries
Robert Plomin

SUMMARY. The links between parent-reported parent-child relationship quality and teacher-reported peer relationship quality were investigated in a sample of biological and adoptive families in the Colorado Adoption Project (CAP; n = 423 children). Data were analyzed using a multilevel model approach. Results indicated that teacher-rated popular-

---

Thomas G. O'Connor is Senior Lecturer in Psychology and Child and Adolescent Psychiatry, Social Genetic and Developmental Psychiatry Research Center, Institute of Psychiatry, 111 Denmark Hill, London, SE5 8AF, UK. Jennifer M. Jenkins is Professor, Institute of Child Study, University of Toronto, 45 Walmer Road, Toronto, Ontario, M5R 2X2, Canada. John Hewitt and John C. DeFries are Professors, Institute for Behavioral Genetics, Campus Box 447, University of Colorado, Boulder, CO 80309. Robert Plomin is Professor, Social Genetic and Developmental Psychiatry Research Center, Institute of Psychiatry, London.

Address correspondence to Thomas G. O'Connor (E-mail: spjwtoc@iop.kcl.ac.uk).

This research was supported by grants HD-10333, HD-18426, and HD36773 from the National Institute of Child Health and Human Development, by grant MH-43899 from the National Institute of Mental Health, the William T. Grant Foundation and grants DA05131 and DA11015.

[Haworth co-indexing entry note]: "Longitudinal Connections Between Parenting and Peer Relationships in Adoptive and Biological Families." O'Connor et al. Co-published simultaneously in *Marriage & Family Review* (The Haworth Press, Inc.) Vol. 33, No. 2/3, 2001, pp. 251-271; and: *Gene-Environment Processes in Social Behaviors and Relationships* (ed: Kirby Deater-Deckard, and Stephen A. Petrill) The Haworth Press, Inc., 2001, pp. 251-271. Single or multiple copies of this article are available for a fee from The Haworth Document Delivery Service [1-800-HAWORTH, 9:00 a.m. - 5:00 p.m. (EST). E-mail address: getinfo@haworthpressinc.com].

ity, but not peer problems, was significantly influenced by genetic factors. Longitudinal analyses indicated that parental report of warmth/ support and child-reported family expressiveness at age 10 years both predicted teacher-rated popularity at age 12 after age 10 popularity was statistically controlled. Also, parental negative control at age 10 years predicted a positive change in peer problems at age 12 years. The effect of parent-child relationship quality on the change in peer relationship quality was obtained in both biological and adoptive families. Finally, evidence of reciprocal influence over time was found for parental negative control and peer problems. *[Article copies available for a fee from The Haworth Document Delivery Service: 1-800-HAWORTH. E-mail address: <getinfo@ haworthpressinc. com> Website: <http://www.HaworthPress.com> © 2001 by The Haworth Press, Inc. All rights reserved.]*

KEYWORDS. Peer relations, parenting, genetics

Understanding how, and indeed if, the effects of family relationships are carried forward to influence children's relationships outside the family is a central task for developmental psychology. The extent to which family relationships predict children's ability to develop positive relationships with peers is an especially important area for research given the strong connections between peer relationships and long-term psychological well-being (Parker & Asher, 1987). The current study examined the links between relationship quality within the family and peer relationship quality in a longitudinal genetically informative study based on data when children were age 10 to 12 years.

FAMILY-PEER LINKS

Evidence supporting a link between quality of relationships in the family and concurrent and subsequent extrafamilial relationships is substantial and derives from several distinct theoretical perspectives (Parke et al., 1989). For instance, several studies demonstrate that the quality of child-parent attachment in infancy and early childhood predicts relationship quality with peers concurrently and longitudinally (Cassidy, Kirsh, Scolton, & Parke, 1996; Moss, Rousseau, Parent, St-Laurent, & Saintonge, 1998; Sroufe, Egeland, & Carlson, 1999). In general, these studies show that, compared to children who were judged to have an insecure attachment relationship with parents, children with a secure attachment relationship are more likely to be rated as popular and accepted by their peers, and to be rated as having more prosocial skills that promote positive peer interactions. Connections between attachment to parents and peers and friendship relationship quality in later childhood and adolescence have also been found (Greenberg, Siegel, & Leitch, 1983; Lieberman, Doyle, & Markiewicz, 1999).

Similarly, research using a social learning approach has established linkages between parenting and peer relationships (Dishion, 1990; Pettit, Dodge, & Brown, 1988; Putallaz, 1987). In this model, the connection between parenting and peer relationships is believed to be mediated by social cognitions and behavioral strategies (e.g., concerning the effectiveness of aggressive behavior) learned from interacting with parents. A related approach proposes that social-cognitive capacities important for positive peer relationships, such as emotional understanding, perspective-taking, and emotional regulation, are developed in the context of the early parent-child relationship and are carried forward or generalized to later social relationships, including those with peers (Carson & Parke, 1996; Dekovic & Janssens, 1992; Dunn, 1992; Parke et al., 1989). Taken together, the above set of findings provides substantial evidence for plausible causal links between the quality of parent-child and peer relationships.

Alternative Hypotheses for the Connection Between Parent and Peer Relationship Quality

Numerous indices of child and adolescent adjustment are under some degree of genetic influence (Rutter, Silberg, Simonoff, & O'Connor, 1999). Most relevant to the current research are the findings suggesting genetic influence on both parent-child and peer relationships. Specifically, genetic influence on several dimensions of parent-child relationships has been suggested by findings from many child-based designs using a range of methods and measures (Lytton, 1977; O'Connor, Hetherington, Reiss, & Plomin, 1995). Genetic mediation of parent-child interaction quality is thought to come about partly via "evocative" processes whereby children evoke warm/supportive, controlling or otherwise negative behavior from parents that amplify, reinforce, or otherwise maintain genetically influenced behavioral characteristics, such as temperament. Thus, for example, children who may be genetically at risk for aggressive behavior may be especially likely to evoke aggressive behavior from others, including parents (Ge et al., 1996; O'Connor, Deater-Deckard, Fulker, Rutter, & Plomin, 1998).

The impact of genetic influences on individual differences may also extend to relationships outside the family. One of the few studies in this area found that siblings' similarity in peer group affiliation was predicted from their genetic resemblance (Manke, McGuire, Reiss, Hetherington, & Plomin, 1995). The implication of those results was that the tendency to actively seek out and ally with a scholastic (or alternatively, athletic or disruptive) peer group is predicted from individual differences in children's personality, cognitive skills and other characteristics that, in turn, are partly genetic influenced. Other studies, using somewhat different measures and methods, also indicate that peer relationship quality and types of affiliation may be partly genetically mediated (Daniels & Plomin, 1985; Leve, this volume).

Given, then, that parent-child relationships and peer relationship are under substantial genetic influence, it could be that the connections between the two are mediated by genetic influences (e.g., on social-cognitive abilities) rather than by relationship influences described above (Rowe, 1989). Although research has yet to examine the genetic mediation of the correlation between parenting and peer relationships, there is a substantial and growing set of studies suggesting that genetic mediation underlies associations between psychosocial risks and individual differences in adjustment (O'Connor, Deater-Deckard, & Plomin, 1998). One reason why the effects of psychosocial risks on children's adjustment may be partly mediated by genetics is that environmental and genetic risks are correlated in development, via active and evocative processes outlined above (Plomin, 1994).

"Passive" Genotype-Environment Correlations

Behavioral genetic research findings illustrate another context in which there is a correlation between genes and environment. In studies of biologically related parents and children, on whom much of our knowledge of development and psychopathology is based, parents provide both genes and environments for their biological children. Accordingly, studies consisting exclusively of biological families are unable to establish a causal connection between qualities of the family *environment* and children's behavioral/emotional adjustment. This latter example of how genes and environment are correlated, which is referred to as "passive" genotype-environment correlation, is a focus of this study.

Failing to account for the fact that genes and environment are correlated in development (such as via "passive" genotype-environment correlations) may lead to overestimating and/or mis-specifying the impact of family factors on children's psychological adjustment. This is because an effect may be thought to be entirely environmentally mediated when in fact genetic mediation is involved. For example, there is some evidence from adoption studies that the correlations between parenting and children's behavioral/emotional adjustment are greater in biological families than in adoptive families (McGue, Sharma, & Benson, 1996), implying that the connection between parenting and behavioral/emotional outcomes is not "purely" environmental in origin and involves some degree of genetic mediation.

We assessed the associations between parenting and peer relationships using data from the Colorado Adoption Project, a longitudinal study of adoptive and biological families. The current study extends previous research in two ways. First, the longitudinal design allowed us to examine whether parent-child relationship quality predicted a *change* in children's peer relationship quality. Second, the genetically informative design provided leverage to investigate whether these connections were environmental in nature or were also partly mediated by genetic influence. In an adoption study design, genetic mediation is suggested if the association between parenting and peer relationships is stronger in biological families compared to adoptive families. If, on the other hand, the connections between parenting and peer relationship quality are equally strong in these family forms, then genetic mediation via passive genotype-environment processes is rejected.

METHOD

Sample and Procedure

The Colorado Adoption Project (CAP; DeFries, Plomin, & Fulker, 1994) is an ongoing, prospective-longitudinal study of adoptive and nonadoptive (i.e., biological) families. Adoptive families were recruited through two large adoption agencies in Colorado from 1975 to 1982. The adopted children's mean age at placement in their adoptive homes was 29 days. Adoptive parents were generally middle-class and well educated. Adoptive mothers' average age was 33 years; adoptive fathers' average age was 34. The number of years of education was 14.7 for mothers and 15.7 for fathers. Over 95% of the adoptive families were Caucasian. Occupational status of the adoptive families, based on National Opinion Research Center ratings of the father's job, was slightly higher than a random sample of families in the Denver metropolitan area based on census data (Plomin, DeFries, & Fulker, 1988). The biological mothers of adopted children were, on average, younger (mean age = 20 years) than adoptive mothers (mean age = 33 years) and completed fewer years of education (12.1 and 14.7 years, respectively). As detailed in Plomin et al. (1988), there was minimal selective placement.

Nonadoptive or biological families were recruited through local hospitals and were matched to the adoptive families on several criteria, including fathers' age, education and occupational status (Plomin et al., 1988). Biological mothers' average age was 30 years; biological fathers' average age was 32. The number of years of education was 14.9 for mothers and 15.6 for fathers. Over 95% of the adoptive families were Caucasian (Plomin et al., 1988).

The initial CAP sample included 245 adoptive and 245 biological families. For the current study, complete data (see measures below) at ages 10 and 12 years were available on 158 adoptive probands and 160 nonadoptive probands. In addition to the probands from adoptive and biological families, the current study also included data from the siblings of CAP probands. There were 96 siblings of probands in adoptive families and 99 siblings of probands in nonadoptive families; in all cases, these siblings were younger than the adoptive or nonadoptive proband in the family. Complete data were available on 52 of the adoptive siblings and 53 of the nonadoptive families. Multilevel model analyses are flexible, and allow for the inclusion of families with one or more children. Consequently, using this methodology we are able to estimate family-level and individual child-level variance without restricting analyses only to the comparatively small number of sibling pairs in adoptive and nonadoptive families. The multilevel analyses presented below were based on 423 children in 337 families.

Data were collected from parents, teachers and the children at ages 10 and 12 years. Parents and children completed the parenting questionnaire at both assessments; teachers' reports were collected via mail. For methodological reasons, notably shared rater bias, we wanted to examine data using different sources of information to examine the concurrent and longitudinal associations between parenting and peer relationship quality. Accordingly, parenting was based on parental report and peer relationship quality was based on teacher report; children provided an index of general family functioning.

Measures

Parenting at Ages 10 and 12 Years. When the children were aged 10 and 12 years, mothers (and in a few cases fathers) completed the Parent Report (Dibble & Cohen, 1974). This measure consists of 48 items that are rated on a 7-point scale (0 = never to 6 = always) and comprise eight positive and eight negative parenting subscales (each containing 3 items). Principal components analyses of the subscales based on oblique rotation suggested one positive and two negative factors. These three factors were robust across assessments. The three factors were Warmth (acceptance, child centeredness, sensitivity, positive involvement), Negative Control (guilt induction, hostility, withdrawal from relationship), and Inconsistency (inconsistent or lax discipline). Each factor was moderately stable across assessments (r = .64 to .70) and the three factors were modestly to moderately inter-correlated within assessments (absolute values of r ranged from r = .13 to r = .37 at age 10 years and from r = .15 to r = .39 at age 12 years).

Teacher Reports of Peer Relationships at Ages 10 and 12 Years. Teacher reports on the Walker Problem Behavior Identification Checklist (WPBIC), a general index of problem behavior, social competence, and peer relationship quality, were collected when the children were 10 and 12 years of age. Teacher responses to the 39 items were rated on a 5-point scale according to how well the item described the child (1 = not at all, 5 = very well). For the current study, these items were subjected to a principal components analysis with varimax rotation. We included the largest possible sample of children at each assessment period. Each assessment for which we factor analyzed teacher report data yielded 6-7 factors (eigen value > 1). Given our longitudinal focus, we were especially interested in factors that were robust across the age period studied. In order for an item (and its associated factor) to be considered we specified that it must load on only one factor at .5 or above and that it loaded on the same factor at each assessment. Using these criteria, we obtained 4 factors. Peer Popularity was defined by items indexing popularity and an ability to get along with others (sample items include "is popular," "has many friends"). Peer Problems was defined by items indexing a general difficulty getting along with age mates, primarily because of aggressive behavior or deficient social skills (sample items include, "starts fights with other kids," "has to do things his/her own way," "hurts others kids' feelings"). Two additional factors describing general personality and social adjustment were identified. Assertiveness was defined by a general tendency to assert and define views positively (sample items include "expressed ideas eagerly," "defends views under pressure"). Finally, maturity was defined as a tendency to act in a responsible manner (sample items include "carries out tasks responsibly," "good organizer, planner"). Internal consistencies for these scales across each assessment period were > .80. Because of our particular interest in peer relationship quality, we focused only on those factors pertaining to peer relationships: Peer Popularity and Peer Problems. These two primary factors accounted for 11%-34% of the variance at ages 10 and 12 years.

The rate of missing data was greater for teachers than for parents or children (most likely because data were collected via post rather than in person). For the age 12 assessment we therefore included data from age 11 years if the age 12 data were missing. This resulted in an additional 104 children across groups. The findings were substantively identical when we used the reduced or extended sample; therefore, analyses reported include the supplemental data at age 11 years.

Family Environment. Children's self-reports on the Family Environment Scale (FES; Moos & Moos, 1981) were collected at ages 10 and 12 years. This measure, which has been used extensively in research, including previous analyses on this sample, provided an additional index of family affective climate in addition to the parenting variables. We included data from the Cohesion, Expressiveness, and Conflict subscales because of our interest in family relationship qualities and affective climate.

Data Analysis. Multilevel modeling (Bryk & Raudenbush, 1992; Goldstein et al., 1998) is designed for hierarchically organized data at a potentially infinite number of levels, such as children within classrooms within schools, or, as in the current study, children within families. Two features of the multilevel model results are highlighted. First, we present the fixed effects associated with the predictor variables. These estimates and standard errors are interpreted as in a regression model; an estimate that is approximately twice its standard error has a significant ($p < .05$) association with peer outcomes. Second, the novel feature of multilevel modeling, the random effects parameters, or the partitioning of variance into each level of the data, is also given. In the "random effects" section, error variance is decomposed into family-level and individual child-level variability. It is important to note that the estimates included in the random effects part of the tables are estimates of variance (with associated standard errors) and should not be interpreted in the same way as the estimates for the fixed effects. Estimates for the fixed and random effects are calculated using a maximum likelihood procedure. The improvement in fit between models can be tested by comparing the change in likelihood ratio with the degrees of freedom equal to the difference in the degrees of freedom of the contrasting models. All analyses were conducted using MLwiN (Goldstein et al., 1998).

For the fixed and random effects parameters, a dummy variable distinguished adoptive and nonadoptive families, coded with adoptive families as the control condition. Thus, the regression coefficients (for fixed effects) and variance estimates (for random effects) indicate the *relative* difference between biological and adoptive families. Given that adoptive siblings are genetically unrelated, significant family-level variance in adoptive families would be explained by "shared environment," or sibling resemblance associated with family membership rather than genetic effects. In biological families, family-level variance could be attributed to shared environment and genes. Therefore the difference in family-level variance (i.e., the extent to which family-level variance is greater in biological compared to adoptive families) is an index of genetic influence. Specifically, genetic influence is suggested by a significantly greater effect of family-level variance in biological families than in adoptive families. The ratio of family-level variance to total variance, or the intraclass correlation, provides a measure of sibling similarity in adoptive and biological families and offers a method of testing genetic hypotheses that complements other approaches.

RESULTS

Descriptive and Correlational Analyses

The correlations among the measures are given in Table 1. There were modest cross-construct correlations across this 2-year period and considerable stability within construct in most cases. It should be noted that the stability in peer variables is considerable despite the fact that different teachers provided data at age 10 and 12 years.

Predictors of Change in Peer Popularity from Age 10 to Age 12 Years

The first aim in the analysis was to distinguish family-level from individual child-level variance in the null model in which no predictor variables were included. Results indicated that 38% of the variation in Popularity was observed at the family-level, with the remaining 62% of variance attributed to factors operating at the individual child level (not

TABLE 1. Correlations Among Peer, Parenting, and Family Relationship Variables and Ages 10 and 12 Years

| | Age 10 Years | | | | | | | |
Age 12 Years	Popularity	Peer Problems	Maternal Warmth	Maternal Neg. Control	Maternal Inconsistency	FES Cohesion	FES Expressive	FES Conflict
Popularity	.48**	-.21**	.13**	-.11*	-.04	.12*	.20**	-.09
Peer Problems	-.21**	.44**	-.13**	.24**	.03	-.08	-.15**	.09
Maternal Warmth	.12*	-.12*	.66**	-.22**	-.28**	.05	.09	-.07
Maternal Negative Control	-.12*	.22**	.23**	.70**	.11*	-.06	-.06	.12*
Maternal Inconsistency	-.14**	.08	-.26**	.05	.64**	-.14**	-.09	.05
FES: Cohesion	.12*	-.10*	.11*	-.11*	-.04	.33**	.21**	-.29**
FES: Expressiveness	.10*	-.10*	.11*	-.13**	.08	.15**	.21**	-.15**
FES: Conflict	-.15**	.12*	.00	.16**	.03	-.27**	-.18**	.40**

Note: Teachers provided information on Popularity and Peer Problems. Mothers reported on Warmth, Negative Control, and Inconsistency, and children reported on family Cohesion, Expressiveness. and Conflict. *p < .05. ** p < .01. n = 423 children.

tabled). This indicates that much of the variation in Peer Popularity is accounted for by factors that operate at the level of the individual child. Nevertheless, a substantial portion can be attributed to factors that operate at the family level (which includes genetics) and make children in the same family similar to one another.

Next, the dummy variable for family type was entered as a fixed effect to examine if Popularity differed for adoptees and nonadoptees. The regression coefficient was nonsignificant, suggesting no mean differences in Popularity between adoptees and nonadoptees. To test the hypothesis of genetic influence, we examined the family-level variation in Peer Popularity in biological and adoptive families separately. In this analysis, the control condition was the variance in adoptive families and we tested the hypothesis that there would be additional family-level variance in biological families. This resulted in a significant improvement in the fit of the model (change, $(\chi^2(1) = 5$, $p < .05$). An examination of the random effects parameters (Table 2, Model 1) indicated more than twice as much family-level variation in Peer Popularity in Biological families (.22 + .32) compared to Adoptive Families (.22). Interestingly, the family-level random effect coefficient for the adoptive families (i.e., the control condition) was just significant, indicating that some of the effects on sibling similarity on Popularity could be attributed to family-level influences.

It is possible to translate these findings on Popularity into an intraclass correlation, an index of sibling similarity. This is computed as the percentage of family-level variance to total variance for Adoptive and Biological families. For adoptive families, the intraclass correlation was .22/(.22 + .66) or .25; for biological families, the intraclass correlation was greater, (.22 + .32)/((.22 + .32) + .66) or .45. No differences were detected between adoptive and biological families in individual child-level variance. That is, individual child-level variation was comparable in adoptive and biological families; therefore, all subsequent models include random effects coefficients only at the family level.

Subsequent models included the age 10 index of Popularity (Model 2), individual child-level variables of child sex and parent-child relationship variables at age 10 (Model 3), and presumed family-level family functioning scales reported by the child at age 10 years (Model 4). Several findings stand out. First, despite marked stability in Popularity across this relatively short time span, several factors predicted change. Increased levels of Popularity were observed for girls, for children with warm and engaged parents at the prior assessment, and children in fami-

TABLE 2. Prediction of Change in Teacher-Reported Peer Popularity at 12 Years: Fixed and Random Effects (SE)

	Model 1	Model 2	Model 3	Model 4
Fixed Effects				
Non-adoptive family	.14 (.11)	.09 (.09)	.13 (.09)	.13 (.09)
Age 10 Popularity		.46 (.04)*	.45 (.04)*	.44 (.04)*
Child Sex			.25 (.08)*	.22 (.08)*
Mother-child Warmth			.014 (.006)*	.012 (.006)*
Mother-child Negative Control			−.002 (.006)	−.004 (.006)
Mother-child Inconsistency			.007 (.008)	.006 (.007)
Family Cohesion				.00 (.02)
Family Expressiveness				.05 (.02)*
Family Conflict				.01 (.011)
Random effects				
Family-level				
Control (Adoptive families)	.22 (.11)*	.17 (.09)	.15 (.08)	.17 (.08)*
Biological families	.32 (.14)*	.23 (.11)*	.24 (.11)*	.29 (.11)*
Individual Child-level				
Control (Adoptive families)	.66 (.10)*	.52 (.07)*	.50 (.07)*	.45 (.07)*
Biological families	---	---	---	---
−2*log(likelihood)	1200	1092	1075	1064

Note: Standard errors are given in parentheses. Family type was dummy coded with Adoptive families as the control condition. Thus, the regression coefficients (for fixed effects) and variance estimates (for random effects) indicate the effect in Biological families relative to Adoptive families. Each model is a significant improvement in fit from the previous model at $p < .05$. $n = 423$.

lies characterized by emotional expressiveness. The prediction of change from prior family and relationship variables is notable, but the effect size was small. Second, variables that significantly predicted variation in Popularity (i.e., their inclusion was associated with reduced variance in bottom half of Table 2) reduced individual child-level variation. This was equally true of the variables assessed at the level of the individual (parent-child relationship quality) and at the level of the family (family expressiveness). Third, further analyses (not tabled) indicated that there were no significant interactions between family type (i.e., adoptive versus biological family) and parenting variables. In other words, the prediction of teacher-rated peer relationship quality from parent reports of parenting was not significantly different in adoptive and biological families. This suggests that passive genotype-environment correlations did not play a mediating role in the prediction of *change* in Popularity from parent-child and family relationship quality. Finally, results from the final model in Table 2 indicate that, even when the predictor variables were included, there remained significantly more family-level variance in biological families compared to adoptive families (Model 4). As a consequence, the ratio of family-level variation to total variation was substantially greater in biological families (.51) than adoptive families (.27). This indicates that more of the variation in Popularity is accounted for by family level factors in biological than adoptive families.

Predictors of Change in Peer Problems from Age 10 to Age 12 Years

Table 3 displays the parallel analyses for predicting age 12 Peer Problems. The pattern of family-level and individual child-level variance differed from that found for Popularity. All of the detectable variation in Peer Problems at age 12 years and *change* of Peer Problems from age 10 to age 12 years was observed at the individual child level, and this was the case when we considered adoptive and biological families together or analyzed family-level variance separately for each group. Thus, all of the variation in the change in Peer Problems between assessments was due to factors operating at the individual child level. Child-level variance did not differ for adoptive and biological families (results not tabled); therefore all models were run without random effects coefficients at the individual child level. Table 3 (Model 2) indicates that there was modest stability in Peer Problems from age 10 to age 12 years. Subsequent models indicated that an increase in Peer Problems between assessments was observed for children whose mothers reported higher levels of negative control at age 10 years (Table 3, Model 3). Child reports on the family environment scales did not predict change in Peer Problems once parenting variables were included in the model (Table 3, Model 4). Nor was the change in Peer Problems associated with the parenting variables significantly different in adoptive and biological families, suggesting no role of passive genotype-environment correlation mediating change (results not tabled).

TABLE 3. Prediction of Change in Teacher-Reported Peer Problems at 12 Years: Fixed and Random Effects (SE)

	Model 1	Model 2	Model 3	Model 4
Fixed Effects				
Non-adoptive family	−.16 (.07)*	−.11 (.06)*	−.13 (.06)*	−.13 (.06)*
Age 10 Peer Problems		.38 (.04)*	.34 (.04)*	.33 (.04)*
Child Sex			−.11 (.06)	−.10 (.06)
Mother-child Warmth			−.005 (.004)	−.004 (.004)
Mother-child Negative Control			.0011 (.004)*	.012 (.004)*
Mother-child Inconsistency			−.001 (.005)	−.001 (.005)
Family Cohesion				.001 (.012)
Family Expressiveness				−.018 (.012)
Family Conflict				−.002 (.008)
Random Effects				
Family-level				
Control (Adoptive families)	.00 (.00)	.00 (.00)	.00 (.00)	.00 (.00)
Biological families	--	--	--	--
Individual Child-level				
Control (Adoptive families)	.46 (.03)*	.37 (.05)*	.36 (.03)*	.35 (.03)*
Biological families	--	--	--	--
−2*log(likelihood)	871	781	763	761

Note: Standard errors are given in parentheses. For the fixed and random effects parameters, family type was dummy coded with Adoptive families as the control condition. Thus, the regression coefficients (for fixed effects) and variance estimates (for random effects) indicate the effect in Biological families relative to Adoptive families. Each model is a significant improvement in fit from the previous model at p < .05. *n* = 423.

Reciprocal Relations Between Parenting and Peer Relationships

A final set of analyses was carried out to test the hypothesis that there was a reciprocal association between parent-child relationship quality and peer relationship quality. For these analyses, which were also conducted using MLwiN, we predicted parental Warmth or Negative Control at age 12 years after first entering the dummy variable distinguishing adoptive and biological families and the prior assessment of parenting at age 10. We first tested the hypothesis that a change in Warm/supportive parenting was predicted by peer Popularity. No evidence of this was found. That is, the age 10 Popularity did not predict age 12 Warm/supportive parenting once the age 10 measure of parenting was statistically controlled. In contrast, an increase in parental Negative Control between ages 10 and 12 was predicted by higher levels of Peer Problems at age 10 (beta = .78 SE = .37, p < .05).

DISCUSSION

Developmentalists agree that a central developmental task of early adolescence is the formation of positive peer relationships, but debate remains about the sources of influence and the mechanisms involved. The current study contributes to this line of research in two ways. Firstly, we examined whether or not change in peer relationship quality in early adolescence was predicted from qualities of the family environment. We found that short-term change in peer relationship quality was predicted from family factors, with warm, positive relationships predicting a positive change in Popularity and parental negative control predicting an increase over time in peer problems. Importantly, the effect was not exclusively from parenting to peer relationship quality, as an increase in parental negative control was predicted by prior peer problems. Secondly, because the research was based on a genetically informative sample of adoptive and biological families, we were able to examine the extent to which the change in peer relationship quality associated with family environment was genetically mediated. The findings indicated that there was genetic influence on teacher-reported Popularity. Nevertheless, the change in peer relationship quality from parenting was equally strong in adoptive and biological parent-child dyads, suggesting no genetic mediation of this link via a "passive" correlation between genes and environment.

Limitations

Before discussing the implications of the findings, we first review some of the study's limitations. The first limitation to note is that the number of families contributing two (or more) children per family was relatively small. Of course, sibling and family studies that use this methodology (O'Connor, Dunn, Jenkins, Pickering, & Rasbash, 2001; Rowe, Almeida, & Jacobson, 1999) will necessarily include comparatively few observations (i.e., children) per family. This is in contrast to the more typical studies using multilevel analysis based on, for example, research in education that may include many children nested within different classrooms. Nevertheless, this is not a problem for the analytic approach and, in any event, we were able to detect significant family-level variance of change in peer popularity even over a relatively short time period. As the current study illustrates, the methodology's potential for testing new hypotheses in developmental psychology and behavioral genetics seems promising and warrants further attention. An additional limitation was that we were unable to specify the mechanisms that underlie the concurrent and longitudinal connections between quality of relationships with parents and peers. There remains a number of alternative hypotheses that require further consideration, such as the role of cognitive processes (e.g., Cassidy et al., 1996).

Given the methodological concerns about rater bias, we used reports from three separate sources, parents, children and teachers. One consequence of this was the loss of data associated particularly with teacher reports. Although the actual rate of missing data per assessment or variable was modest, the cumulative effect was more marked. Offsetting these potential limitations are a number of strengths, the most important of which are the prediction of intra-individual change and the use of different reporters for parent-child and peer relationship quality.

Prediction of Change

Although the opportunity to test behavioral genetic hypotheses was the most novel feature of the study, the findings also contribute to the extant research on family and peer linkages (Parke et al., 1989). The central finding in this regard, that change in peer relationships from parent-child relationship quality, is notable given that comparatively few of the existing studies linking parenting and peer relationships consider change over time. The magnitude of the effect was small, however. In addition, these findings extend and complement prior research in demonstrating connections in late childhood and early adolescence and in showing such connections for both positive and conflict/aggressiveness dimensions (Vuchinich, Bank, & Patterson, 1992).

There has long been an interest in, and empirical support for, the role of peer relationships in children's social and personality development and psychopathology (Sullivan, 1953; Cicchetti & Bukowski, 1995; Cowen, Pedersen, Babigian, Izzo, & Trost, 1973). A particularly important finding in this line of research concerns the connections among the "social worlds" of children (see Hartup, 1979; Parke et al., 1989; Sroufe et al., 1999). That is, rather than set in opposition the influence of parents and peers on children's development, as is sometimes suggested (Harris, 1995), it is important to consider the links between these sources of influence and the fact that they do not operate in isolation. Indeed, in the case of negative parental control and peer problems, there was some suggestion for reciprocal influences between parent-child relationships and peer relationships.

Multilevel Analyses and Behavioral Genetic Hypotheses

One of the main lessons from this research that synthesizes developmental psychology with behavioral genetics is that genetic influence extends well beyond the more traditional confines of intelligence and temperament or personality. In fact, a surprisingly large number of developmentally important outcomes appear to show some degree of genetic influence–although the explanations for the nature of the genetic effect are often lacking. Findings from this report extend prior studies (Daniels & Plomin, 1985; Manke et al., 1995) in showing that individual differences in teacher-rated popularity are partly genetically influenced, although no such effect was obtained for peer problems. In the context of multilevel models, this interpretation is based on the greater intraclass correlation in biological families than adoptive families. The greater sibling resemblance in the former than in the latter is likely mediated by the greater genetic similarity in siblings in biological families. It is possible that there are other factors besides, or in addition to, genetics that operate on a family-wide basis to make siblings in biological families more similar in popularity than those in adoptive siblings. Indeed, this is the basis for some criticisms of the adoption study design. Nevertheless, sibling similarity in biological families remained significantly greater on popularity even after we statistically controlled for parent-child relationships and family affective climate. In fact, there was no appreciable change in intraclass correlations in adoptive and biological families in the null and full regression models. Thus, it is at least the case that these family process factors did not account for greater resemblance of biological siblings.

In addition to testing genetic influence on outcomes, developmental behavioral genetics also considers the degree to which phenotypic associations may be genetically mediated. There are, for example, a number of illustrations that parenting-child psychopathology associations are mediated by genetics, in the form of passive or evocative genotype-environment correlations (Ge et al., 1996; McGue et al., 1996). Accordingly, just as the association between parenting and child psychopathology may be partly genetically mediated, we considered the possibility that the associations between parenting and *changes* in peer relationships were genetically mediated. Results indicated that this was not the case. Specifically, the prediction of change in peer popularity or problems from parent-child relationship quality was comparable in adoptive and biological families. That is, there was no interaction between the parenting measure and sibling type in predicting change. The absence of genetic mediation should be considered in the context of the fact that most studies showing genetic mediation are based on cross-sectional analyses rather than analyses of intra-individual change. Connections between parent-child relationship quality and *changes* in peer relationships over time are instead mediated by environmental factors, which were specifically defined in this research as parent-child relationship quality and family affective climate.

In addition to testing genetic hypotheses, the use of multilevel modeling provided additional insights into the nature of environmental influence. For example, the results indicated that much of the variance in peer outcomes was attributable to factors operating at the level of the individual child rather than factors that operate on a family-wide basis. This was indicated by the decomposition of variance in the null models and further suggested by the drop in child-level variance when predictor variables were added into the regression model. Importantly, the drop in individual child-level variance was found regardless of whether the predictor variable was thought to index individual-specific risks such as parenting or family-level processes such as the family environment scales. In other words, both kinds of processes appeared to be explaining variation at the individual or child-specific level–regardless of the level (i.e., child or family) at which the process was measured. The same lesson can be derived from a wealth of behavioral genetics data (Plomin & Daniels, 1987), which uses a different methodology to support the same conclusion. The implication is that there is a need to focus on independent child-specific risks and risk mechanisms; the role of family-level risk processes seems comparatively less salient. Nevertheless, there was significant family-level variance in popularity in adoptive families, suggesting that the effects of some environmental factors may be to make siblings similar to one another.

There is growing appreciation for the carrying-forward of the effects of family relationships in development. This is now well established in terms of extrafamilial relationships such as peers in cross-sectional and short-term longitudinal studies. More recently, long-term longitudinal studies demonstrate that the effects of parent-child interaction quality in childhood and adolescence predict adult interpersonal relationship quality with an intimate partner (Conger, Cui, Bryant, & Elder, 2000) and with adult peers (Allen, Hauser, O'Connor, & Bell, 2002). Despite the long-term nature of these connections, the reasons for the links are still not well understood. Stable personality dispositions, attachment style and social cognitions, and other factors may all be involved in maintaining these links. Progress in understanding the nature of these links will require further synthesis of concepts and methods from developmental psychology and behavioral genetics.

REFERENCES

Allen, J. P., Hauser, S. T., O'Connor, T. G., & Bell, K. B. (2002). Prediction of peer-rated adult hostility from autonomy struggles in adolescent-family interactions. *Development and Psychopathology, 14*, 123-137.

Bryk, A. S., & Raudenbush, S. W. (1992). *Hierarchical linear models.* Newbury Park, CA: Sage.

Carson, J. L., & Parke, R. D. (1996). Reciprocal negative affect in parent-child interactions and children's peer competency. *Child Development, 67*, 2217-2226.

Cassidy, J., Kirsh, S. J., Scolton, K. L., & Parke, R. D. (1996). Attachment and representations of peers. *Developmental Psychology, 32*, 892-904.

Cicchetti, D., & Bukowski, W. M. (1995). Developmental processes in peer relations and psychopathology. *Development and Psychopathology, 7*, 587-589.

Conger, R. D., Cui, M., Bryant, C. M., & Elder, G. H. Jr. (2000). Competence in early adult romantic relationships: A developmental perspective on family influences. *Journal of Personality and Social Psychology, 79*, 224-37.

Cowen, E. L., Pedersen, A., Babigian, H., Izzo, L. D., & Trost, M. A. (1973). Long-term follow-up of early detected vulnerable children. *Journal of Consulting and Clinical Psychology, 41*, 438-446.

Daniels, D., & Plomin, R. (1985). Differential experience of siblings in the same family. *Developmental Psychology, 21*, 747-760.

DeFries, J. C., Plomin, R., & Fulker, D. W. (1994). *Nature and nurture during middle childhood.* Cambridge, MA: Blackwell.

Dekovic, M., & Janssens, J. M. (1992). Parents' child-rearing style and child's sociometric status. *Developmental Psychology, 28*, 925-932.

Dibble, E., & Cohen, D. J. (1974). Companion instrument for measuring children's competence and parental style. *Archives of General Psychiatry, 30*, 805-815.

Dishion, T. (1990). The family ecology of boys' peer relations in middle childhood. *Child Development, 61*, 874-892.

Dunn, J. (1992). *Young children's close relationships: Beyond attachment.* Newbury Park, CA: Sage.

Ge, X., Conger, R. D., Cadoret, R. J., Neiderhiser, J. M., Yates, W., Troughton, E., & Stewart, M. A. (1996). The developmental interface between nature and nurture: A mutual influence model of child antisocial behavior and parent behaviors. *Developmental Psychology, 32*, 574-589.

Goldstein, H., Rasbash, J., Plewis, I., Draper, D., Browne, W., Yang, M., Woodhouse, G., & Healy, M. (1998). *A user's guide to MLwiN*. Institute of Education: London.

Greenberg, M. T., Siegel, J. M., & Leitch, C. J. (1983). The nature and importance of attachment relationships to parents and peers during adolescence. *Journal of Youth and Adolescence, 12*, 373-386.

Harris, J. R. (1995). Where is the child's environment? A group socialization theory of development. *Developmental Review, 102*, 458-489.

Hartup, W. W. (1979). The social worlds of childhood. *American Psychologist, 34*, 944-950.

Lieberman, M., Doyle, A. B., & Markiewicz, D. (1999). Developmental patterns in security of attachment to mother and father in late childhood and early adolescence: Associations with peer relations. *Child Development, 70*, 202-213.

Lytton, H. (1977). Do parents create, or respond to, differences in twins? *Developmental Psychology, 13*, 456-459.

Manke, B., McGuire, S., Reiss, D., Hetherington, E. M., & Plomin, R. (1995). Genetic contributions to children's extrafamilial social interactions: Teachers, best friends, and peers. *Social Development, 4*, 238-256.

McGue, M., Sharma, A., & Benson, P. (1996). The effect of common rearing on adolescent adjustment: Evidence from a U.S. adoption cohort. *Developmental Psychology, 32*, 604-613.

Moos, R. H., & Moos, B. S. (1981). *Family Environment Scale manual*. Palo Alto, CA: Consulting Psychologists Press.

Moss, E., Rousseau, D., Parent, S., St-Laurant, D., & Saintonge, J. (1998). Correlates of attachment at school age: Maternal reported stress, mother-child interaction, and behavior problems. *Child Development, 69*, 1390-1405.

O'Connor, T. G., Deater-Deckard, K., Fulker, D. W., Rutter, M., & Plomin, R. (1998). Genotype-environment correlations in late childhood and early adolescence: Antisocial behavioral problems and coercive parenting. *Developmental Psychology, 34*, 970-981.

O'Connor, T. G., Deater-Deckard, K., & Plomin, R. (1998). Foundations of behavioral genetics research: Implications for clinical psychology. In C. E. Walker (Ed.), *Comprehensive Clinical Psychology, Vol. 1 Foundations* (pp. 87-114). Oxford: Elsevier Sciences.

O'Connor, T. G., Dunn, J., Jenkins, J. M., Pickering, K., & Rasbash, J. (2001). Family settings and children's adjustment: Differential adjustment within and across families. *British Journal of Psychiatry, 179*, 110-115.

O'Connor, T. G., Hetherington, E. M., Reiss, D., & Plomin, R. (1995). A twin-sibling study of observed parent-adolescent interactions. *Child Development, 66*, 812-829.

Parke, R. D., MacDonald, K. B., Burks, V. M., Carson, J., Bhavnagri, N., Barth, J. M., & Beitel, A. (1989). Family and peer systems: In search of the linkages. In K. Kreppner & R.M. Lerner (Eds.), *Family systems and life-span development* (pp. 56-92). Hillsdale, NJ: Erlbaum.

Parker, J. G., & Asher, S. R. (1987). Peer relations and later personal adjustment: Are low accepted children at risk? *Psychological Bulletin, 102*, 357-389.

Pettit, G. S., Dodge, K. A., & Brown, M. M. (1988). Early family experience, social problem solving patterns, and children's social competence. *Child Development, 59*, 107-120.

Plomin, R. (1994). *Genetics and experience.* Thousand Oaks, CA: Sage.

Plomin, R., & Daniels, D. (1987). Why are children in the same family so different from one another? *Behavioral and Brain Sciences, 10*, 1-16.

Plomin, R., DeFries, J. C., & Fulker, D. W. (1988). *Nature and nurture in early childhood.* New York: Cambridge University Press.

Putallaz, M. (1987). Maternal behavior and children's sociometric status. *Child Development, 58*, 324-340.

Rowe, D. C. (1989). Families and peers: Another look at the nature-nurture question. In T.J. Berndt & G.W. Ladd (Eds.), *Peer relationships in child development* (pp. 174-199). New York: Wiley.

Rowe, D. C., Almeida D. M., & Jacobson K. C. (1999). School context and genetic influences on aggression in adolescence. *Psychological Science, 10*, 277-280.

Rutter, M., Silberg, J., O'Connor, T., & Simonoff, E. (1999). Genetics and child psychiatry: II. Empirical research findings. *Journal of Child Psychology and Psychiatry, 40*, 19-55.

Sroufe, L. A., Egeland, B., & Carlson, E. A. (1999). One social world. In W. A. Collins & B. Laursen (Eds.), *Minnesota symposium on child psychology: Vol. 30* (pp. 241-262). Hillsdale, NJ: Erlbaum.

Sullivan, H. S. (1953). *Interpersonal theory of psychiatry.* New York: Norton.

Vuchinich, S., Bank, L., & Patterson, G. R. (1992). Parenting, peers, and the stability of antisocial behavior in preadolescent boys. *Developmental Psychology, 28*, 510-521.

Nature and Nurture in the Family

Robert Plomin

Kathryn Asbury

SUMMARY. Behavioral genetic research provides the best evidence we have for the importance of the environment. Heritabilities rarely exceed 50%, which means that, on average, half of the variance for behavioral dimensions and disorders is not genetic in origin. However, no one needed to be convinced that environmental factors contribute importantly to behavior. The two findings that stand out as the most important from behavioral genetics both involve nurture rather than nature. This volume reviews these findings and lays out the agenda for future research in this area. Some of the ideas described in these papers may be unfamiliar to marriage and family researchers, in part because they are so novel and revolutionary. Nonetheless this volume will reward careful reading as it represents an exciting glimpse of the future of research in this area. *[Article copies available for a fee from The Haworth Document Delivery Service: 1-800-HAWORTH. E-mail address: <getinfo@haworthpressinc.com> Website: <http://www.HaworthPress.com> © 2001 by The Haworth Press, Inc. All rights reserved.]*

KEYWORDS. Behavioral genetics, nonshared environment, gene-environment correlation

---

Robert Plomin is Deputy Director of the Social, Genetic and Developmental Psychiatry Research Center, and Kathryn Asbury is a PhD student, both at the Institute of Psychiatry, King's College London, 111 Denmark Hill, London SE5 8AF, UK.

Preparation of this paper was supported in part by a program grant from the UK Medical Research Council.

[Haworth co-indexing entry note]: "Nature and Nurture in the Family." Plomin, Robert, and Kathryn Asbury. Co-published simultaneously in *Marriage & Family Review* (The Haworth Press, Inc.) Vol. 33, No. 2/3, 2001, pp. 273-281; and: *Gene-Environment Processes in Social Behaviors and Relationships* (ed: Kirby Deater-Deckard, and Stephen A. Petrill) The Haworth Press, Inc., 2001, pp. 273-281. Single or multiple copies of this article are available for a fee from The Haworth Document Delivery Service [1-800-HAWORTH, 9:00 a.m. - 5:00 p.m. (EST). E-mail address: getinfo@haworthpressinc.com].

This truly special volume illustrates the point that behavioral genetics is telling us much more about nurture (environment) than about nature (genetics). The message about nature is important but embarrassingly simple: Everything shows genetic influence. Instead of asking if genetics is important, it is now more appropriate to ask whether there is a behavioral dimension or disorder that does *not* show genetic influence.

Behavioral genetic research provides the best evidence we have for the importance of the environment. Heritabilities rarely exceed 50%, which means that, on average, half of the variance for behavioral dimensions and disorders is not genetic in origin. However, no one needed to be convinced that environmental factors contribute importantly to behavior. The two findings that stand out as the most important from behavioral genetics both involve nurture rather than nature. This volume reviews these findings and lays out the agenda for future research in this area. Some of the ideas described in these papers may be unfamiliar to marriage and family researchers, in part because they are so novel and revolutionary. Nonetheless this volume will reward careful reading as it represents an exciting glimpse of the future of research in this area.

NONSHARED ENVIRONMENT

The first major finding from behavioral genetics relates to nonshared environment, which is discussed in the paper by McGuire. For nearly all behaviors, the environmental factors that affect behavioral development do not make children growing up in the same family similar to one another. Only genes run in families. When some or all family members are not genetically related, as in adoptive or stepfamilies, there is little resemblance.

Why two children in the same family are so different is the phenomenon to be explained. This phenomenon was identified by what McGuire calls 'outcome-oriented' research which is behavioral genetic research that focuses on outcomes such as adjustment or personality. This research shows that family members do not share most environmental influences. But the research does not identify the experiential sources of nonshared environment. The 'experience-oriented' perspective begins to address this issue by focusing on differences in experience among family members, especially siblings. Sibling differences in experience, whether objectively or subjectively assessed, are considerable. However, the most important step in the nonshared environment program of research is to ask whether differences in experience are associated with differences in outcome. Sibling differences in experience are surely more important if they make a difference to developmental outcomes. And they do. The meta-analysis described by McGuire (Turkheimer & Waldron, 2000) shows that the proportion of total variance accounted for in adjustment, personality and cognitive outcomes was .01 for family constellation, .02 for differential parental behavior, .02 for differential sibling interaction, and .05 for differential peer or teacher interaction. Moreover, these effects are largely independent because aggregate measures of differential environments account for 13 percent of the total variance.

More dramatically, using composite measures of family environment and adolescent outcome, the Nonshared Environment in Adolescent Development (NEAD) project found strikingly high associations between differential sibling experiences and differential sibling outcomes (Reiss, Neiderhiser, Hetherington, & Plomin, 2000). If NEAD had only included siblings and not made use of genetically sensitive designs such as MZ and DZ twins, and half siblings and genetically unrelated siblings in 'blended' families, it would have been tempting to declare victory in the quest for nonshared environment. However, when the genetically sensitive design was brought to bear, it was clear that most of what appeared to be nonshared environment was in fact mediated genetically. Results from the genetically sensitive design indicate that parents respond to genetically driven differences in their children rather than creating those differences environmentally.

THE NATURE OF NURTURE

This brings us to the second major behavioral genetic discovery about the environment, which has been called the nature of nurture, a topic discussed in the paper by Towers et al. Genetic factors not only affect the usual behavioral traits that have been studied in behavioral genetic research such as cognitive abilities and disabilities, personality and psychopathology, but they can also affect behaviors in social settings, including parent, sibling, friend and peer relationships. Towers et al. describe new research showing genetic influence on parenting, and their own new study that shows genetic influence (and no influence of shared environment) on marital relationships, which are so often used as environmental measures in child development research. The paper by Lemery and Goldsmith reports a new study of sibling relationships that suggests genetic influence on sibling conflict but not sibling cooperation. This seems to illustrate a more general trend toward finding greater genetic influence on negative rather than positive aspects of family relationships (Reiss et al., 2000). This paper finds that the sibling relationship is a source of nonshared environmental influence on difficult temperament.

Genetic influences have been found to contribute to nearly all measures of psychological experience that have been studied because, like family environment and family relationships, measures of psychological experiences involve behavior (Plomin, 1994). Psychological environments are not independent of a person, like the weather. Consciously or unconsciously, people select, modify and create their own environments, and they do this in part to construct environments that correlate with their genetic propensities.

Finding genetic influence on measures of the environment means that one day we will be able to find polymorphisms in DNA that account for these genetic influences, a fascinating possibility mentioned in Eley's paper on molecular genetics. The paper by Towers et al. describes the first study of family relationships that is currently collecting DNA for this purpose. Although talking about molecular genetics in relation to family relationships might seem far-fetched, the molecular genetics revolution is proceeding at a pace unseen ever before in science, as indicated by the publication early in 2001 of the working draft of the human genome sequence and identification of half of the 3 million DNA variants in the genome sequence. Using specific genes as markers of genetic influence on experience will greatly sharpen investigations of the developmental interface between nature and nurture. Moreover, when DNA markers are found that are associated with family relationships and family environment, it will be easy for marriage and family researchers to incorporate the DNA markers into their research, just like other risk indicators, without the need for special samples like twins or adoptees. Collecting DNA does not require blood—just rubbing the inside of the cheek with a Q-tip can provide enough DNA to genotype hundreds of DNA markers.

If genetics contributes to experiences as well as to behavioral outcomes, it is not surprising to find that associations between experiences and behavioral outcomes are often mediated in part by genetic factors. For example, as shown by the NEAD project, associations between parenting and children's adjustment are mediated substantially by genetic factors. The paper by Towers et al. discusses an important distinction between child-based genetic studies of parenting like NEAD and parent-based studies like the new Twin Mothers project. The paper by

Riggins-Caspers and Cadoret describes the excellent work of the Iowa group that uses an adoption design to explore independent and interactive contributions of genetic risk (based on biological parents' alcoholism and antisocial behavior) and environmental risk (parenting and psychopathology of adoptive parents) as they relate to adolescent adoptees' problem behaviors. Stepfamilies are increasingly common and often include half siblings whose genetic similarity is 25% because they share only one parent in common as well as full siblings whose genetic similarity is 50%. The paper by Deater-Deckard, Dunn, O'Connor, Davies and Golding describes the results of a new study that uses this genetically sensitive design to explore family processes. They also call for more studies of this type, which complement twin and adoption studies but also provide researchers with larger and more representative samples than twin and adoption studies.

Genetic mediation of associations between family environment and children's outcomes can be viewed as genotype-environment correlations (Plomin, 1994). As Towers et al. point out, and also Riggins-Caspers and Cadoret, this is a useful way of exploring the interface between nature and nurture in family relationships that have typically been conceptualized as environmental, and of identifying those environmental factors that can most usefully be targeted for intervention. The reason why the NEAD book (Reiss et al., 2000) is called *The Relationship Code* is that genotype-environment correlation restores to the family some of its influence lost to nonshared environment in the sense that genetic propensities are expressed in the family environment. Genotype-environment correlation is neither genetic nor environmental–it is both.

NONSHARED ENVIRONMENT AGAIN

The discovery of genetic mediation of associations between family relationships and children's outcomes is important but brings us back to nonshared environment. If associations between sibling differences in family experiences and sibling differences in outcome are mediated by genetic factors, what environmental factors are responsible for nonshared environment? The bottom line is that nonshared environment is where the environmental action is in terms of behavioral outcomes. There is a lot to be done before we understand either what it is or how it operates.

So what is it? As McGuire points out, future research in this area will consider developmental issues and the role of context but the most obvious direction for research is to consider experiences outside the family. In retrospect, as Harris has pointedly pointed out, it seems a bit daft to have looked for differential experiences of siblings in the family because siblings live in the same family. However, in defense of NEAD, investigating the family environment as a first step in the quest for nonshared environment allowed NEAD to capitalize on the huge research effort that has gone into assessing family environment. Indeed, it turns out that siblings experience very different family environments and that these differential experiences are strongly associated with differential outcomes. The problem is that genetic mediation largely accounts for these associations. Despite the results of NEAD, families can still be mined profitably for possible sources of shared and nonshared environmental influence in domains other than psychopathology such as cognitive and reading development which is the target of a new adoption study described by Deater-Deckard, Petrill and Wilkerson. Another novel approach to family environment is to assess round-robin interactions, which can be analyzed using a version of the Social Relations Model with a genetic extension, as discussed by Manke and Pike.

As suggested by Harris and the meta-analysis mentioned earlier, extrafamilial factors seem good candidates for nonshared environments and have not been studied nearly as much as family environment. Harris pins her hopes on peers, and three papers in this volume focus on peers. One of these papers directly tests the extent to which peers are responsible for nonshared environment independent of genetic factors. Using observational twin data, Leve reports that peer interaction shows substantial nonshared environmental influence in line with Harris' theory and in contrast to results for parent-child interaction. The other two papers are not directly relevant to the issue of peers as nonshared environmental influence but provide glimpses of the interesting research that will come from studying peers in genetically sensitive designs. Gilson, Hunt and Rowe use siblings to compare two theories on the origins of peer similarity—social homogamy and peer selection—and find different results for delinquency and verbal intelligence. O'Connor, Hewitt, DeFries and Plomin use data from the Colorado Adoption Project and find some genetic influence on teacher-rated peer popularity but not on peer problems. They also show that parent-child relationships can predict changes in peer relationship quality free of genetic mediation.

There is much more to be learned about peers and other extrafamilial experiences as sources of nonshared environment. However, nonshared environment appears to be present early in life, long before children experience peers or other extra familial influences, which implies that such extrafamilial factors cannot completely account for nonshared environment.

No matter how difficult it may be to find specific nonshared environmental factors, it should be emphasized that nonshared experience is how the effective environment operates to create individual differences (Plomin, Asbury, & Dunn, 2001). More research is needed to understand these mechanisms. The price of admission to research on nonshared environment is to study more than one child per family. This is not difficult because more than 80 percent of families have more than one child and it gives researchers two children for little more than the price of one. Siblings add synergistically to any research design: You can study whatever you were going to study in the traditional between-family way and then look at the data from a within-family perspective, which can clarify the interpretation of any between-family findings. Any study of siblings can be used to investigate the first two steps in the program of research mentioned earlier, documenting differential experiences and the association between differential experiences and differential outcomes. Longitudinal studies can also begin to address the third step, interpreting the direction of effects for the association between for example parenting (X) and child outcome (Y)–whether X causes Y or Y causes X or a third factor is responsible for the association between X and Y. Behavioral genetic studies of twins or adoptees are useful to investigate genetics as a possible 'third factor' mediating such associations.

We also need to consider the gloomy (for researchers, though not necessarily for the rest of humanity) prospect that chance contributes to nonshared environment in the sense of random noise, idiosyncratic experiences or the subtle interplay of a concatenation of events. Francis Galton, the founder of behavioral genetics, suggested that nonshared environment is largely due to chance: "The whimsical effects of chance in producing stable results are common enough." Sounding a bit like a fortune cookie, he observed that "Tangled strings variously twitched, soon get themselves into tight knots."

Two overlooked findings from behavioral genetic research point to the importance of chance (Plomin et al., 2001). The first is multivariate genetic research showing that nonshared environmental influences on one trait such as depression are independent of nonshared environmental influences on other traits such as antisocial behavior. The second finding involves the application of multivariate genetic analysis to longitudinal data, which assesses the genetic and environmental origins of age-to-age change and continuity. Longitudinal genetic analyses indicate that nonshared environmental influences are age specific for psychopathology, personality, and cognitive abilities. In NEAD some stability for nonshared environment was found over a 3-year interval in adolescence for depression and antisocial behavior, although the majority of nonshared environment was unstable (Reiss et al., 2000). That is, nonshared environmental influences at one age are largely different from nonshared environmental influences at another age.

What environmental processes other than chance could explain these two findings? Nonetheless, our view is that chance is the null hypothesis. Systematic sources of nonshared environment need to be thoroughly examined before we shrug our shoulders and call it chance. Chance might only be a label for our current ignorance to identify the processes by which children–even pairs of identical twins–growing up in the same family come to be so different. To some extent children might make their own luck. Identical twins may use their identical genetic material and shared environment as the springboard into different but parallel universes. Studying the nonshared environment will tell us something about how and why they do this, thereby enhancing our understanding of the origins of individual differences and the clinical, academic and philosophical implications of such differences.

A larger implication of nonshared environment is that it provides an excellent example that some of the most important questions for genetic research involve the environment and some of the most important questions for environmental research involve genetics. Genetic research will profit if it includes sophisticated measures of the environment, environmental research will benefit from the use of genetic designs, and understanding of development will be advanced by collaboration between geneticists and environmentalists. These are ways in which behavioral scientists are putting the nature-nurture controversy behind them and bringing nature and nurture together in the study of development in order to understand the processes by which genotypes become phenotypes.

REFERENCES

Galton, F. (1889). *Natural inheritance*. London: Macmillan.

Harris, J. R. (1998). *The nurture assumption: Why children turn out the way they do.* New York: The Free Press.

Plomin, R. (1994). *Genetics and experience: The interplay between nature and nurture.* Thousand Oaks, California: Sage Publications.

Plomin, R., Asbury, K., & Dunn, J. (2001). Why are children in the same family so different? Nonshared environment a decade later. *Canadian Journal of Psychiatry, 46,* 225-233.

Reiss, D., Neiderhiser, J.M., Hetherington, E.M. & Plomin, R. (2000). *The relationship code: Deciphering genetic and social patterns in adolescent development.* Cambridge, MA: Harvard University Press.

Turkheimer, E, & Waldron, M. (2000). Nonshared environment: A theoretical, methodological, and quantitative review. *Psychological Bulletin, 126,* 78-108.

Cults and the Family, edited by Florence Kaslow, PhD, and Marvin B. Sussman, PhD (Vol. 4, No. 3/4, 1982). *"Enlightens not only the professional but the lay reader as well. It provides support and understanding for families . . . gives insight and . . . enables parents, friends, and loved ones to better understand what happens when one joins a cult." (The Family Psychologist)*

Family Medicine: A New Approach to Health Care, edited by Betty Cogswell and Marvin B. Sussman (Vol. 4, No. 1/2, 1982). *The history, rationale, and the continuing developments in this medical specialty all in one readable volume.*

Marriage and the Family: Current Critical Issues, edited by Marvin B. Sussman (Vol. 1, No. 1, 1979). *Covers pluralistic family forms, family violence, never married persons, dual career families, the "roleless" role (widowhood), and non-marital, heterosexual cohabitation.*

Index